React.js 16
从入门到实战

王金柱 著

清华大学出版社
北京

内 容 简 介

本书是一本从实战出发，学习React.js 16框架及其相关技术栈的入门与实践书籍。全书内容翔实、重点突出、代码丰富、通俗易懂，涵盖了React.js 16框架开发的方方面面。

全书共分为16章，包括React基础开发、React JSX、组件Components、Props参数、State状态和生命周期、事件处理、条件渲染、列表、表单、组合与继承、状态提升、Refs、Router路由与Redux扩展等相关知识内容。另外，为了突出本书项目实战的特性，有针对性地基于React框架及其技术栈讲解了5个实际项目应用，可以帮助读者深入掌握React的开发技巧。

本书用于学习React.js 16编程，全书内容简洁、代码精练、重点突出、实例丰富，能够帮助初学者快速掌握React开发方法。同时，对于前端开发人员提高服务器端脚本语言的开发技术水平有非常大的帮助。

本书封面贴有清华大学出版社防伪标签，无标签者不得销售。
版权所有，侵权必究。举报：010-62782989，beiqinquan@tup.tsinghua.edu.cn。

图书在版编目（CIP）数据

React.js 16从入门到实战 / 王金柱著.—北京：清华大学出版社，2020.1（2021.12重印）
（Web前端技术丛书）
ISBN 978-7-302-54543-9

Ⅰ.①R… Ⅱ.①王… Ⅲ.①JAVA语言－程序设计 Ⅳ.①TP312.8

中国版本图书馆CIP数据核字（2019）第290388号

责任编辑：夏毓彦
封面设计：王　翔
责任校对：闫秀华
责任印制：杨　艳

出版发行：清华大学出版社
网　　址：http://www.tup.com.cn，http://www.wqbook.com
地　　址：北京清华大学学研大厦A座　　邮　编：100084
社 总 机：010-62770175　　邮　购：010-62786544
投稿与读者服务：010-62776969，c-service@tup.tsinghua.edu.cn
质量反馈：010-62772015，zhiliang@tup.tsinghua.edu.cn

印 装 者：三河市铭诚印务有限公司
经　　销：全国新华书店
开　　本：190mm×260mm　　印　张：22　　字　数：563千字
版　　次：2020年1月第1版　　印　次：2021年12月第2次印刷
定　　价：69.00元

产品编号：080568-01

前　言

读懂本书

React 异军突起

　　React 框架自诞生伊始就备受瞩目，一切皆源自其强大的背景。React 框架最初是社交网络巨头 Facebook（脸书）公司的一个内部项目，设计目标是用来架构 Instagram 网站的。Instagram 就是大名鼎鼎的、用于图片分享的社交应用，用户可通过 Instagram 随时随地将抓拍的图片上传到移动终端设备（手机、平板电脑等）上彼此分享。

　　本书注重代码实践，为读者全面深入地讲解了针对 React 16 开发的技术栈。全书百余个代码实例给读者带来的不仅仅是全面的基础知识，更是为读者提供了设计简洁、开发高效的实际项目应用。可以说，这是一本学习 React 16 框架开发技术的高效手册。

　　React 支持跨终端、多平台的开发，无论是传统的 PC 端、还是现今正流行的移动端设备，又或是 Windows 系统、Android 系统和 iOS 系统，均是 React 可以发挥威力的舞台。可以说，今天 React 框架的迅速崛起是由其内在的技术特性所决定的。

React 技术领先

　　React 框架的设计初衷主要是用于构建 UI，而构建 UI 的核心思想就是封装组件。组件维护自身的状态和 UI，每当状态发生改变时就会自动重新渲染组件自身，而不需要通过反复查找 DOM 元素后再重新渲染整个组件了。

　　同时，React 框架支持传递多种类型的参数，比如：代码声明、动态变量，甚至是可交互的应用组件。因此，UI 渲染方式既可以通过传统的静态 HTML DOM 元素，也可以通过传递动态变量，甚至是通过整个可交互的组件来完成的。

　　本书中关于以上这些技术内容均有详细介绍，通过具体的代码实例帮助读者学习和掌握这些知识点的原理及使用方法，真正实现了理论与实践相结合的学习方法。

本书真的适合你吗？

　　本书涵盖了绝大部分关于 React 16 基础及进阶的内容，将知识点与应用实例相结合，通过大量的代码实例帮助读者快速掌握 React 16 框架的编程技巧，并应用到实践开发之中。本书通过这种学以致用的方式来增强读者的阅读兴趣，对于无论是基础内容或提高内容，相信读者都可以从中获益。

本书内容安排

本书共 16 章，各章节针对不同的 React 16 知识点进行了详细的介绍：

第 1 章主要介绍了关于 React 16 框架安装、环境搭建和开发工具使用方面的内容，旨在帮助读者快速掌握 React 开发的方法。

第 2 章主要介绍了关于 React JSX 方面的知识，JSX 其实就是 JavaScript XML 的缩写，具有支持自定义属性和很好扩展性的特点，是 React 框架所独有的内置语法，因此建议设计人员使用 JSX 方式来实现 UI 中的虚拟 DOM。

第 3 章主要介绍了 React 组件和 Props 参数方面的内容。React 组件可以将 UI 切分成一些独立的、可复用的部件，这样有助于设计人员专注于构建每一个单独的部件。React 组件通过 Props 可以接收任意的输入值，因此 Props 也可以理解为参数的概念。

第 4 章主要介绍了 React State（状态）和生命周期方面的知识。React 将组件看成是一个状态机（State Machines），通过其内部定义的状态（State）与生命周期（Lifecycle）实现与用户的交互，并维持组件不同的状态。

第 5 章主要介绍了 React 事件处理方面的内容。React 框架的事件处理机制与 JavaScript 的 DOM 元素事件类似，但二者在语法上是略有不同的。通过本章的学习，可以帮助读者理解 React 事件处理的使用方法。

第 6 章主要介绍了关于 React 条件渲染方面的内容。在 React 框架中，设计人员可以创建不同的组件来封装各种业务需求，然后依据需求的不同状态，仅仅渲染组件中对应状态下的局部内容。

第 7 章主要介绍了关于 React 列表的知识。具体包括如何创建、转化和使用列表，以及如何通过 Key 来识别元素改变的操作。

第 8 章主要介绍了关于 React 表单的内容。具体包括如何在 React 框架中使用表单（Form），以及与 HTML 表单的异同。React 表单中的 DOM 元素与 React 框架中的其他 DOM 元素有所不同，因为表单元素需要保留一些内部状态（State）。

第 9 章主要介绍了关于 React 组合与继承方面的知识。React 官方推荐设计人员在实际项目中尽量使用组合模式、而非继承来实现组件的代码重用。因此在本章中将对这两种模式进行一些基本的阐述。

第 10 章主要介绍了关于 React 状态提升方面的内容。在 React 框架中，当多个组件需要反映相同的变化数据，这时建议将共享状态提升到最近的共同父组件中去，这就是所谓的"状态提升"。

第 11 章主要介绍了关于 React 扩展方面的内容。所涉及的内容包括 Node、Babel、Webpack、Browserify、React Router 和单页面，以及 Redux 等。这些知识点或是与 React 环境相关、或者与 React 扩展相关，都是与 React 设计开发息息相关的内容。

第 12~16 章针对 React 及其技术栈专门开发了 5 个实际的项目应用，具体包括 React、React-Router 和 Redux 等方面的内容，尽量帮助读者在实践中学习和掌握 React 框架开发 Web 应用的方法和技巧。

本书特点

（1）本书从最简单的、最通用的 React 代码实例出发，抛开枯燥的纯理论知识介绍，通过实例讲解的方式帮助读者快速学习 React 开发技巧。

（2）本书内容涵盖了 React 框架及其技术栈开发所涉及的绝大部分知识点，将这些内容整合到一起可以系统地了解和掌握这门语言的全貌，为介入大型 Web 项目的开发做了很好的铺垫。

（3）本书对于实例中的知识难点做出了详细的分析，能够帮助读者有针对性地提高 React 编程开发技巧，并且通过多个实际的项目应用，尽力帮助读者掌握 React 框架开发所涉及方方面面的内容。

（4）本书在 React 及其相关知识点上按照类别进行合理的划分，全部的代码实例都是独立的，读者可以从头开始阅读，也可以从中间开始阅读，不会影响学习进度。

（5）本书代码遵循重构原理，避免代码污染，真心希望读者能写出优秀的、简洁的、可维护的代码。

源代码下载

本书示例源代码下载地址请扫描右边二维码获得。

如果下载有问题，请直接联系 booksaga@163.com 解决问题，邮件主题为"React.js 16 从入门到实战"。

本书读者

- React 框架学习初学者
- Node.js 框架学习初学者
- JavaScript 开发初学者和前端开发初学者
- 由 JavaScript 向 React 框架转型的开发人员
- IT 培训学校前端实践课程的学生
- 高等院校前端实践课程的学生

关于封面照片

封面照片由蜂鸟网的摄影家 ptwkzj 先生友情提供，在此表示衷心感谢。

作 者
2020 年 1 月

目　录

第 1 章　React 简介 1
 1.1　React 概述 1
 1.2　React 优势 1
 1.3　第一个 React 应用 2
 1.4　React 脚手架 5
  1.4.1　关于 React 脚手架 5
  1.4.2　Node.js 与 npm 6
  1.4.3　Webpack 模块打包器 7
  1.4.4　ES6 和 Babel 7
  1.4.5　使用 create-react-app 脚手架开发第一个应用 8
  1.4.6　React 项目架构 9
 1.5　React 虚拟 DOM 10
 1.6　JSX 简单入门 14
 1.7　React 渲染机制 16

第 2 章　React JSX 19
 2.1　JSX 介绍 19
 2.2　JSX 独立文件 19
 2.3　JSX 算术表达式 21
 2.4　JSX 条件表达式 23
 2.5　JSX 嵌入表达式 24
 2.6　JSX 对象表达式 26
 2.7　JSX 函数表达式 28
 2.8　JSX 增强函数表达式 29
 2.9　JSX 数组表达式 32
 2.10　JSX 样式表达式 33
 2.11　JSX 注释表达式 35

第 3 章　React 组件与 Props 37

- 3.1　React 组件介绍 37
- 3.2　React 函数组件 38
- 3.3　React 类组件 39
- 3.4　React 组合组件 42
- 3.5　React Props 介绍 45
- 3.6　React Props 应用 47
- 3.7　React Props 只读性 49
- 3.8　React Props 默认值 52
- 3.9　React 组件切分与提取 54

第 4 章　React State 与生命周期 63

- 4.1　React State 介绍 63
- 4.2　关于定时器时钟的思考 65
- 4.3　开始封装时钟 UI 66
- 4.4　实现时钟 UI 的自动更新 68
- 4.5　自我更新的时钟 UI 类 69
- 4.6　在时钟组件类中引入 State 状态 72
- 4.7　React 生命周期介绍 76
- 4.8　在时钟组件类中使用生命周期 77
- 4.9　正确的使用 State 状态 79
- 4.10　自顶向下的数据流 84

第 5 章　React 事件处理 92

- 5.1　React 事件介绍 92
- 5.2　React 单击事件 94
- 5.3　React 阻止事件默认行为 97
- 5.4　React 类的事件处理方法 99
- 5.5　绑定回调方法的其他方式 101
- 5.6　在事件处理方法中传递参数 105
- 5.7　实战：开关按钮 109
- 5.8　React 文本框事件 113
- 5.9　实战：水温监控控件 117

第 6 章　React 条件渲染 122

- 6.1　React 条件渲染介绍 122
- 6.2　元素变量的条件渲染 125

6.3　逻辑"与"运算符的条件渲染 133
6.4　逻辑"或"运算符的条件渲染 135
6.5　三元逻辑表达式的条件渲染 137
6.6　实战：改进登录组件 139
6.7　阻止组件渲染 142

第 7 章　React 列表与 Key 145

7.1　React 列表介绍 145
7.2　基础列表组件 147
7.3　多级列表组件 149
7.4　React Key 介绍 152
7.5　React Key 使用 155
7.6　React 通过 Key 提取组件 157
7.7　React Key 局部唯一性 162
7.8　React Key 有效性 164

第 8 章　React 表单 168

8.1　React 表单介绍 168
8.2　表单受控组件 171
8.3　强制转换大写字母 177
8.4　校验手机号码格式 181
8.5　格式化序列号 187
8.6　文本域关键字 190
8.7　下拉列表受控组件 194
8.8　处理多个输入 197
8.9　React 表单提交操作 200
8.10　React 表单提交服务器 205
8.11　受控组件与非受控组件 210

第 9 章　组合与继承 215

9.1　组合与继承概述 215
9.2　定义组件容器 215
9.3　定义子组件 216
9.4　自定义 Props 属性 221
9.5　特例关系组合 223
9.6　类组合方式确认框 225

第 10 章　状态提升 229

10.1　设计构想 229
10.2　实现水温监控功能 230
10.3　加入第二个水温输入框 233
10.4　同步二个水温输入框 237
10.5　将水温"状态共享" 240
10.6　将水温"状态提升" 244
10.7　实现水温同步换算 249

第 11 章　React 扩展 255

11.1　Node.js 与 React 255
11.2　Babel 与 React 256
11.3　Webpack 模块打包器 257
11.4　Node+Babel+Webpack 搭建 React 环境 261
11.5　Browserify 模块打包器 268
11.6　React Router 与单页面应用 271
11.7　Redux 与 React 273

第 12 章　实战 1：基于 React + Redux 实现计数器应用 279

12.1　设计思想 279
12.2　计数器应用页面 280
12.3　主入口模块 281
12.4　视图模块 282
12.5　Action 定义 285
12.6　Reducer 设计 285
12.7　计数器应用测试 286

第 13 章　实战 2：基于 React+Redux 实现计算器应用 289

13.1　设计思想 289
13.2　计算器应用页面 290
13.3　主入口模块 291
13.4　视图模块 292
13.5　Action 定义 296
13.6　Reducer 设计 297
13.7　计算器应用测试 299

第 14 章 实战 3：基于 Provider 容器组件重构计算器应用 301

- 14.1 设计思想 301
- 14.2 主入口模块 303
- 14.3 App 组件 303
- 14.4 视图模块 305
- 14.5 Action 定义 309
- 14.6 Reducer 设计 309
- 14.7 重构的计算器应用测试 312

第 15 章 实战 4：基于 Redux 实现任务管理器应用 313

- 15.1 设计思想 313
- 15.2 任务管理器应用页面 314
- 15.3 主入口模块 315
- 15.4 App 组件 315
- 15.5 AddTodo 组件 316
- 15.6 VisibleTodoList 组件 317
- 15.7 Footer 组件 320
- 15.8 Action 定义 322
- 15.9 Reducer 设计 323
- 15.10 任务管理器应用测试 325

第 16 章 实战 5：基于 React+Router+Redux 的网站架构 328

- 16.1 设计思想 328
- 16.2 网站架构应用页面 329
- 16.3 主入口模块 330
- 16.4 App 组件 330
- 16.5 Reducer 设计 332
- 16.6 视图组件 335
- 16.7 Action 定义 338
- 16.8 网站架构应用测试 339

第 1 章 React 简介

React 是一款目前十分流行的开源前端 Web 框架,在 Web 前端应用设计领域拥有很强的知名度。本章作为全书开篇,将详细向读者介绍关于 React 框架的基础知识及入门应用,帮助读者快速熟悉 React 应用开发的方法和流程。

1.1 React 概述

React 框架自诞生伊始就备受瞩目,一切皆源自其强大的背景。React 框架最初是社交网络巨头 Facebook(脸书)公司的一个内部项目,设计目标是用来架构 Instagram 网站的。Instagram 就是大名鼎鼎的、用于图片分享的社交应用,用户可通过 Instagram 随时随地将抓拍的图片在移动终端设备(手机、平板电脑等)上彼此分享。至于 React 框架与 Instagram 之间,有一段很曲折的故事。

Instagram 其实最初是一家独立公司,于 2012 年被 Facebook 公司收购。Facebook 在考虑为 Instagram 设计 UI 时,对市面上绝大部分很成熟的 JavaScript MVC 前端框架均不太感冒。于是乎,React 就如同大多数的前端框架的诞生一样,被 Facebook 单独开发出来,专门用于设计 Instagram。因此,React 框架的设计思想很独特、视角很新奇,是很具有革命性的一款产品。

目前,React 框架已经被越来越多的设计人员所关注和使用,有一种说法认为它很可能称为未来 Web 开发的主流工具。由于 React 框架的大受欢迎,导致后来项目体量越滚越大,已经从最早的 UI 引擎渐渐演变成了一套覆盖前后端的 Web App 解决方案。React 框架凭借其良好的性能优势、简洁的代码逻辑和庞大的受众群体,已经成为越来越多开发人员进行 Web App 应用开发的首选框架。

React 框架的官方网址为 https://reactjs.org/。

1.2 React 优势

React 框架的设计初衷主要是用于构建 UI,而构建 UI 的核心思想就是封装组件。组件维护自身的状态和 UI,每当状态发生改变时,就会自动重新渲染组件自身,而不需要通过反复查找 DOM 元素后再重新渲染整个组件了。

同时，React 框架支持传递多种类型的参数，如代码声明、动态变量或者是可交互的应用组件。因此，UI 渲染方式既可以通过传统的静态 HTML DOM 元素，也可以通过传递动态变量，还可以通过整个可交互的组件来完成。

下面简单概括一下 React 框架的主要优点：

- 声明式设计：React 采用声明范式，可以轻松描述应用。
- 高效：React 通过对 DOM 的模拟，最大限度地减少与 DOM 的交互。
- 灵活：React 可以与已知的库或框架很好地配合。
- 使用 JSX 语法：JSX 是 JavaScript 语法的扩展，可以极大地提高 JS 运行效率。
- 组件复用：通过 React 构建组件使得代码易于复用，可在大型项目应用开发中发挥优势。
- 单向响应的数据流：React 实现了单向响应的数据流，减少了重复代码，比传统数据绑定方式更简单。

另外，在由原生 React 框架所衍生的 React Native 项目发展过程中，有设计人员希望通过用编写 Web App 的方式去编写 Native App。该方式如能最终实现工业化，相信未来的互联网行业会被重塑。因为，设计人员只需要编写一次 UI，就能生成同时运行在服务器、PC 浏览器和移动终端 App（手机、平板电脑等）。

1.3 第一个 React 应用

本节开始介绍第一个使用 React 框架开发 Web 前端应用，让读者一识 React 应用的庐山真面目。React 应用开发涉及的内容比较广泛，我们还是从最简单、最基本的"Hello World"开始介绍。简单来讲，就是如何将传统意义的 HTML 网页内容，以 React 框架渲染的方式来实现。

首先，就是如何安装和使用 React 框架。React 框架的安装和使用很简单，可以直接通过 CDN 方式获取 React 和 ReactDOM 的 UMD 版本引用，具体如下：

```
<script crossorigin src="https://unpkg.com/react@16/umd/react.development.js" />
<script crossorigin src="https://unpkg.com/react-dom@16/umd/react-dom.development.js" />
```

不过，上述版本仅用于开发环境，不适合生产环境。React 压缩和优化之后的生产环境版本链接如下：

```
<script crossorigin src="https://unpkg.com/react@16/umd/react.production.min.js" />
```

```
<script crossorigin
src="https://unpkg.com/react-dom@16/umd/react-dom.production.min.js" />
```

此外,还要引入 Babel 编译器所需的库文件,具体如下:

```
<!-- Don't use this in production: -->
<script src="https://unpkg.com/babel-standalone@6.15.0/babel.min.js">
</script>
```

> **注 意**
>
> Babel 库文件在生产环境下也是不建议使用的。

关于上面以 CDN 方式引入的一组库文件(react.js、react-dom.js 和 babel.js 这 3 个脚本文件,文件名为泛指),具体描述如下:

- react.js 是 React 框架的核心库。
- react-dom.min.js 提供与 DOM 相关的功能。
- babel.min.js 由 Babel 编译器提供,可以将 ES6 代码转为 ES5 代码,这样就能在不支持 ES6 的浏览器上执行 React 代码(请注意在生产环境中不建议使用)。

下面,开始第一个使用 React 框架实现"Hello World"应用的代码实例,具体如下:

【代码 1-1】(详见源代码目录 ch01-react-helloworld.html 文件)

```
01  <!DOCTYPE html>
02  <html>
03    <head>
04      <meta charset="UTF-8" />
05      <title>Hello World</title>
06      <script src="https://unpkg.com/react@16/umd/react.development.js"></script>
07      <script src="https://unpkg.com/react-dom@16/umd/react-dom.development.js"></script>
08      <!-- Don't use this in production: -->
09      <script src="https://unpkg.com/babel-standalone@6.15.0/babel.min.js"></script>
10    </head>
11    <body>
12      <div id="root"></div>
13      <script type="text/babel">
14        ReactDOM.render(
15          <h1>Hello, world!</h1>,
16          document.getElementById('root')
17        );
18      </script>
```

```
19    </body>
20  </html>
```

关于【代码 1-1】的说明：

- 第 06 行、07 行和第 09 行代码，以 CDN 方式分别引用了 React 框架所需的三个库文件（react.development.js、react-dom.development.js 和 babel.min.js）。
- 第 12 行代码通过<div id="root">标签元素定义了一个层，用于显示通过 React 框架渲染的文本内容。
- 第 14~17 行代码通过调用 React DOM 对象的 render()方法来渲染元素。
- 第 14 行代码定义了要引入的元素节点<h1>。
- 第 15 行代码获取了页面中要渲染的元素节点<div id="root">。
- 然后，第 14~17 行代码通过 ReactDOM.render()方法将<h1>元素节点渲染到页面的层<div id="root">元素节点中。

关于 ReactDOM.render()方法的语法格式如下：

```
ReactDOM.render(element, container[, callback]);
```

语法说明：

- element 参数：必需，表示渲染的源对象（元素或组件）。
- container 参数：必需，表示渲染的目标对象（元素或组件）。
- callback 参数：可选，用于定义回调方法。

对于【代码 1-1】中所使用渲染方法的方式（见第 14~17 行代码），看上去很直观，不过感觉逻辑性不太强。

下面，我们尝试将上述代码以如下的方式进行改写，或许会比较容易理解了，具体如下：

【代码 1-2】（详见源代码目录 ch01-react-ele-helloworld.html 文件）

```
01  <script type="text/babel">
02      const h1 = (<h1>Hello, world!</h1>);
03      var root = document.getElementById('root');
04      ReactDOM.render(h1, root);
05  </script>
```

关于【代码 1-2】的说明：

- 第 02 行代码通过 const 关键字定义了一个常量(h1)，描述了要引入的元素节点<h1>。由此可见，在 React 框架中变量的使用非常灵活，可以将元素节点直接定义为变量形式来使用。
- 第 03 行代码获取了页面中要渲染的元素节点<div id="root">，保存在变量（root）中。
- 第 04 行代码调用 React DOM 对象的 render()方法，将 h1 元素节点渲染到 root 元素节点中。

可以看到，改写后的【代码1-2】逻辑性比较好，读者应该能够很容易分清渲染的源对象和目标对象，以及渲染方式。

下面分别使用 Firefox 浏览器运行测试【代码1-1】和【代码1-2】定义的 HTML 网页，二者的具体效果（相同的）如图1.1所示。页面中成功显示了通过 React 框架渲染出的文本内容（Hello, world!）。

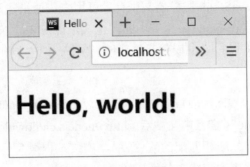

图 1.1　React 实现"Hello world"

1.4　React 脚手架

本节介绍 React 脚手架方面的内容，脚手架是学习 React 开发非常重要的部分，其所涵盖的内容也非常广泛。因此，在介绍一些相关技术点时会简明扼要一些，更深入地了解就需要读者自行完成了。

1.4.1　关于 React 脚手架

如今，Web 前端应用几乎都拥有非常复杂的项目架构，绝非如【代码1-1】那样的单页面文件就可以完成的。React 项目开发也是如此，比较流行的做法是采用 Webpack + ES6 模式来搭建项目架构，然后通过打包方式发布 Web 应用。一般业内就称这个项目架构为"脚手架"，使用这个名词概念确实很形象，就如同盖房子时必须使用的脚手架一样。

上面关于 React 项目架构的描述比较简单抽象，而且还有一些支撑开发 React 应用的关键技术没有描述出来。其实，涉及 React 项目开发的技术点有很多，比如：Node.js 框架、npm 包管理工具、Webpack 模块打包器、ES6（ECMAScript 6），以及 Babel 编译器等。这些内容会在后面章节中有一个大致的介绍，以帮助读者理解 React 脚手架的原理和使用方法。

虽然设计人员通过自己搭建 React 脚手架也是一个很好的学习过程，但有时候难免会遇到问题，或者是做着重复造轮子的工作。因此，设计出一个具有良好兼容性的 React 脚手架，通过自动化方式进行配置，并生成实际项目框架的方式逐渐成为主流。迄今为止，React 脚手架的成熟产品网上有很多，不过最著名的还是 Facebook 自己推出的"create-react-app"脚手架。在本书的项目实践中，也是通过"create-react-app"脚手架来构建 React 应用的。

1.4.2 Node.js 与 npm

如果读者想使用"create-react-app"脚手架进行开发，必须先进行安装。安装"create-react-app"脚手架最常用的方式就是通过 npm 工具进行，那么 npm 工具是什么呢？

简单来讲，npm 就是基于 Node.js 框架的包管理工具。而 Node.js 框架则是基于著名的 Google V8 引擎所开发的 JavaScript 服务器端平台，可用来快速搭建易于扩展的 Web 应用。

npm 包管理工具是绑定在 Node.js 框架中的，也就是说安装好 Node.js 框架后，npm 工具也自动安装完成了。下面以 Windows 操作系统环境为例，介绍安装和部署 Node.js 框架和 npm 包管理工具的方法。

（1）用户可以直接从 Node.js 的官方网站（https://nodejs.org/en/download/）来获取安装包文件，我们也可以通过 Node.js 官方中文站点（http://nodejs.cn/download/）进行下载。下载时，请注意选择与自己电脑操作系统相匹配的软件版本。安装过程很简单，根据自身需要按步骤操作"下一步"按钮，直到程序安装成功就可以了。

（2）判断 Node 程序是否安装成功。判断方法很简单，直接在命令行控制台中通过"node -v"命令查看版本号就可以。如果能够显示出版本号，就表示安装成功了，具体如图 1.2 所示。

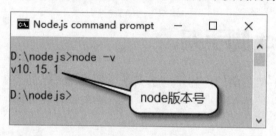

图 1.2　查看 Node.js 版本号

（3）判断 npm 包管理工具是否同时安装成功。判断方法同样很简单，直接在命令行控制台中通过"npm -v"命令查看版本号。如果能够显示出版本号，就表示 npm 工具也安装成功了，具体如图 1.3 所示。

图 1.3　查看 npm 版本号

npm 包管理工具安装成功后，就可以通过该工具安装各种第三方开发包了，也包括"create-react-app"脚手架的安装。

1.4.3　Webpack 模块打包器

前面提到 Webpack 这项技术，那么，Webpack 具体是什么呢？简单来讲，Webpack 是一项模块打包器技术，通过分析项目结构找到 JavaScript（ECMAScript）模块以及其他一些浏览器不能直接运行的扩展语言或资源，将其转换并打包为合适的格式以供浏览器解析。

Webpack 的工作方式可以概括如下：首先将项目作为一个整体来看，通过一个给定的主文件或入口文件（如 index.js），Webpack 将从这个文件开始查找项目的所有依赖文件，并使用加载器（loaders）进行处理，最后打包为一个（或多个）可通过浏览器解析的 JavaScript 脚本文件。

安装 Webpack 很简单，通过 npm 包管理工具在终端命令行操作就可以完成，具体命令如下：

```
npm install -g webpack            // 全局安装方式
npm install --save-dev webpack    // 项目目录安装方式
```

> **说　明**
>
> 设计人员在使用"create-react-app"脚手架的过程中，直观上是感受不到 Webpack 模块打包器存在的，但在底层逻辑上它确实在发挥关键作用。

1.4.4　ES6 和 Babel

ES6 就是指 ECMAScript 6 的缩写。ECMAScript 是一种由 Ecma 国际（欧洲计算机制造商协会前身）标准化的脚本程序设计语言，也可以认为就是读者所熟知的 JavaScript 脚本语言。

不过，JavaScript 与 ECMAScript 还是有所区别的。JavaScript 是先于 ECMAScript 开发出来并付诸实际应用的，ECMAScript 是在 JavaScript 逐渐成熟并成为事实上的脚本语言标准后，经由 Ecma 国际标准化（基于 ECMA-262 标准）的脚本语言。因此，JavaScript 可以理解为是 ECMAScript 标准的实现和扩展。

ES6（ECMAScript 6）是 ECMAScript 系列标准中具有重大意义的一版，它使得 ECMAScript 正式成为国际标准，并被称为下一代的脚本语言。由于 ES6 是在 2015 年正式发布的，因此 ES6 也称为 ECMAScript 2015，且后续的版本开始按照所发布年份命名了。

介绍完 ECMAScript，接着介绍 Babel。Babel 其实是一个工具链，主要用于将基于 ECMAScript 2015+ 标准编写的脚本代码转换为向后兼容 JavaScript 语法的脚本代码，以便能够运行在当前或旧版本的浏览器中。因此，Babel 也可以当作下一代 JavaScript 脚本语言的编译器。

ES6 与 Babel 的组合，是开发 React 应用的重要基础。设计人员在使用"create-react-app"脚手架的过程中，直观上也是感受不到 Babel 存在的，同样是在底层逻辑发挥重要作用。

1.4.5 使用 create-react-app 脚手架开发第一个应用

现在，我们开始介绍如何使用"create-react-app"脚手架开发第一个应用。如果要使用"create-react-app"脚手架，在早些时候就需要先安装该脚手架。不过，自 node v6+和 npm v5.2+版本以后，npm 新引入一条 npx 命令，支持通过直接临时安装"create-react-app"脚手架来创建应用的方式。关于这两种安装方式，下面都会具体介绍。

通过传统全局方式安装"create-react-app"脚手架的方法，具体操作命令参考如下：

```
npm install -g create-react-app         // 全局安装方式
```

在"create-react-app"脚手架安装完毕后，选定好自己的项目目录，通过 create-react-app 命令创建应用，具体命令如下：

```
create-react-app ch01-react-app-demo        // 创建 React 项目应用
```

如果是通过 npx 命令方式，安装"create-react-app"脚手架与创建项目应用是一步完成的，具体操作命令参考如下：

```
npx create-react-app ch01-react-app-demo        // npx 方式
```

通过 npx 方式安装的"create-react-app"脚手架，其实是临时的安装包，项目创建完成后会自动删除该安装包。

以上这两种通过"create-react-app"脚手架创建 React 项目的方式，虽然过程有所区别，但最终效果是一样的。图 1.4 所示是 React 项目创建完成后的提示信息。

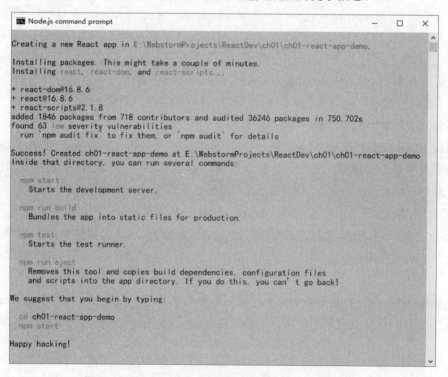

图 1.4 通过"create-react-app"脚手架创建 React 项目

React 项目创建完成后，通过"npm start"命令就可以启动项目了，具体如下：

```
cd ch01-react-app-demo        // 进入项目目录
npm start                     // 启动 React 项目
```

如果通过"npm start"命令启动 React 项目成功，控制台终端中会给出正确的提示信息，如图 1.5 所示。

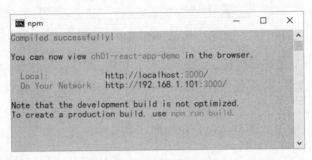

图 1.5　启动 React 项目

通过浏览器访问地址（http://localhost:3000）就可以看到项目启动后的内容了，具体效果如图 1.6 所示。它展示的是一个默认的 React 项目运行后的效果，读者应该感到很振奋人心吧。

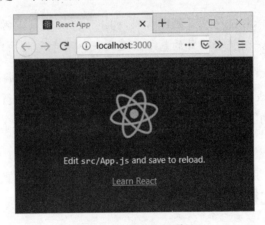

图 1.6　访问 React 项目

1.4.6　React 项目架构

在这个项目应用介绍中，笔者不打算给出任何程序代码，就是让读者了解一下通过"create-react-app"脚手架创建 React 项目的过程。下面看一下通过脚手架创建 React 项目的架构，如图 1.7 所示。

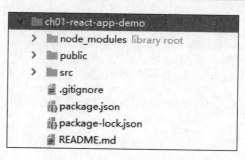

图 1.7　React 脚手架项目架构

图 1.7 展示的是一个最基本的 React 脚手架项目架构。需要说明的是，随着"create-react-app"脚手架应用的不断迭代更新，目录结构和文件名称可能会有修订，不过基本原理还是会保持延续。

图 1.7 是一个默认脚手架项目的目录和文件，关于各个目录和文件的功能作用描述如下：

- node_modules 目录：是通过 npm 工具安装的所有依赖模块，是整个项目的基础核心。
- public 目录：该目录下的 index.html 页面文件是整个项目的入口页面，相当于网站主页。
- src 目录：用于保存整个项目的源代码，其中的 index.js 是源代码入口。
- package.json 文件：整个项目的基本配置文件，用于定义项目的基本信息。

以上就是一个最基本的、默认的"create-react-app"脚手架项目中，关于目录和文件的基本介绍。整个"create-react-app"脚手架项目的架构高度可配置，设计人员可通过自定义方式实现更为复杂和高级的功能。

1.5　React 虚拟 DOM

React 框架的核心优势之一，就是支持创建虚拟 DOM 来提高页面性能。那么，什么是虚拟 DOM 呢？其实，虚拟 DOM 这个概念很早就被提出来了，是相对于实际 DOM 而言的。

设计人员在设计传统 HTML 网页的 UI 时，都会在页面中定义若干的 DOM 元素，这些 DOM 元素是所谓的实际 DOM。通常，页面中的实际 DOM 负责承载着外观表现和数据变化，任何外观形式的改变或数据信息的更新都要反馈到 UI 上，都是需要通过操作实际 DOM 来实现的。

于是，问题也就自然出现了。对于复杂的页面 UI 而言，往往会定义大量的实际 DOM。频繁地操作大量实际 DOM，往往会带来访问性能的严重下降，用户体验也会随之变差，这些都是设计人员所不希望看到的。因此，React 框架专门针对这个现象引入了虚拟 DOM 机制，以避免频繁的 DOM 操作带来的性能下降问题。

React DOM 类似于一种将相关的实际 DOM 组合在一起的集合，是有区别于传统概念上的

DOM 元素的，如果将其理解为 DOM 组件应该更为贴切。因此，React 框架将 React DOM 称为虚拟 DOM。

为了更好地让读者理解虚拟 DOM 的概念，我们分别以两种 DOM 方式来实现效果相同的页面，通过实际代码的区别来进行对比。有趣的是，在 HTML DOM 和 React DOM 语法中均定义了一个 createElement()方法用来创建元素。当然，HTML DOM 语法中的 createElement()方法创建的是一个实际 DOM，而 React DOM 中的 createElement()方法创建的却是一个虚拟 DOM。

下面先看一个通过 HTML DOM 语法中的 createElement()方法创建实际 DOM 的代码实例，具体如下：

【代码 1-3】（详见源代码目录 ch01-js-createElement.html 文件）

```html
01  <!DOCTYPE html>
02  <html>
03  <head>
04      <meta charset="UTF-8" />
05      <title>JavaScript createElement()</title>
06  </head>
07  <body>
08  <!-- 添加文档主体内容 -->
09  <div id='id-div-react'></div>
10  <script type="text/javascript">
11      // TODO: get div
12      var divReact = document.getElementById('id-div-react');
13      // TODO: JavaScript DOM
14      const jsSpan = document.createElement('span');
15      const jsH3 = document.createElement('h3');
16      jsH3.innerText = "JavaScript DOM";
17      const jsP = document.createElement('p');
18      jsP.innerText = "Create dom by JavaScript's createElement() func.";
19      jsSpan.appendChild(jsH3);
20      jsSpan.appendChild(jsP);
21      divReact.appendChild(jsSpan);
22  </script>
23  </body>
24  </html>
```

关于【代码 1-3】的说明：

- 第 09 行代码通过<div id='id-div-react'>标签元素定义了一个层，用于显示通过 JavaScript 创建实际 DOM 的容器。
- 第 14~21 行代码用于在层容器<div id='id-div-react'>中创建实际 DOM 节点，具体内容如下：

➢ 第 14 行代码调用 Document 对象的 createElement()方法，创建了一个<span>元素节点。
➢ 第 15~16 行代码调用 Document 对象的 createElement()方法，创建了一个<h3>元素节点，并定义了文本内容。
➢ 第 17~18 行代码再次调用 Document 对象的 createElement()方法，创建了一个<p>元素节点，同样定义了文本内容。
➢ 第 19~20 行代码分别调用 appendChild()方法，将<h3>和<p>元素节点填充进<span>元素节点内。
➢ 第 21 行代码再次调用 appendChild()方法，将<span>元素节点填充进层容器<div id='id-div-react'>内，从而实现创建实际 DOM 的操作。

下面使用 Firefox 浏览器运行测试该 HTML 网页，具体效果如图 1.8 所示。

图 1.8　JavaScript 创建实际 DOM

下面再看一个使用 React DOM 语法中的 createElement()方法，以虚拟 DOM 方式实现的代码实例，具体如下：

【代码 1-4】（详见源代码目录 ch01-react-createElement.html 文件）

```
01  <!DOCTYPE html>
02  <html>
03    <head>
04      <meta charset="UTF-8" />
05      <title>React createElement()</title>
06      <script src="https://unpkg.com/react@16/umd/react.development.js"></script>
07      <script src="https://unpkg.com/react-dom@16/umd/react-dom.development.js"></script>
08      <!-- Don't use this in production: -->
09      <script src="https://unpkg.com/babel-standalone@6.15.0/babel.min.js"></script>
10    </head>
```

```
11    <body>
12      <!-- 添加文档主体内容 -->
13      <div id='id-div-react'></div>
14      <script type="text/babel">
15        // TODO: get div
16        var divReact = document.getElementById('id-div-react');
17        // TODO: React DOM
18        const reactH3 = React.createElement("h3", {}, "React DOM");
19        const reactP = React.createElement("p", {}, "Create virtual DOM by React's createElement() func.");
20        const reactSpan = React.createElement("span", {}, reactH3, reactP);
21        ReactDOM.render(reactSpan, divReact);
22      </script>
23    </body>
24  </html>
```

关于【代码 1-4】的说明：

- 第 13 行代码通过<div id='id-div-react'>标签元素定义了一个层，用于显示通过 React 创建虚拟 DOM 的容器。
- 第 14～22 行定义的脚本代码用于实现 React 虚拟 DOM，具体内容如下：
  - 需要注意一点，第 14 行代码中的<script>标签元素内，type 属性类型定义为 "text/babel"，表示内部的脚本代码要使用 React 框架进行解析。
  - 第 16 行代码先获取了层<div id='id-div-react'>容器对象（divReact）。
  - 第 18 行和第 19 行代码调用 React DOM 对象的 createElement()方法，分别创建了一个<h3>元素节点（reactH3）和一个<p>元素节点（reactP），并相应定义了文本内容。
  - 第 20 行代码调用 React DOM 对象的 createElement()方法，创建了一个<span>元素节点（reactSpan），然后将刚刚创建的<h3>元素节点（reactH3）和<p>元素节点（reactP）填充进去。
  - 第 21 行代码调用 React DOM 对象的 render()方法，将<span>元素节点（reactSpan）渲染到层<div id='id-div-react'>容器对象（divReact）中进行显示。

下面使用 Firefox 浏览器运行测试该 HTML 网页，具体效果如图 1.9 所示。

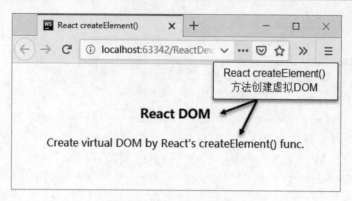

图 1.9　React 创建虚拟 DOM

如图 1.8 和图 1.9 中的箭头和标识所示，通过 HTML DOM 创建的实际 DOM 与通过 React DOM 渲染出来的虚拟 DOM，二者的显示效果是完全相同的。

1.6　JSX 简单入门

在前文中介绍了如何通过 React DOM 的 createElement()方法创建虚拟 DOM，并将创建的虚拟 DOM 渲染到页面中的过程。但是，Facebook 的 React 研发团队还是觉得不太满意，最终开发出来一种专属于 React 框架的 JavaScript 语法扩展——JSX。

所谓 JSX 其实就是 JavaScript XML 的缩写，直译过来就是基于 JavaScript 的 XML。JSX 看起来似乎是一种 XML 格式，其本质仍旧是一种 JavaScript 语言，只不过是将 JavaScript 脚本代码写成 XML 样式。其实，在【代码 1-2】中的第 02 行代码就已经使用到了 JSX 语法，具体如下：

```
02    const h1 = (<h1>Hello, world!</h1>);
```

这里的<h1>标签元素就是通过 JSX 方式定义的，严格讲就是一个虚拟 DOM。为了详细介绍 React 框架中 JSX 方式的使用方法，下面看一个具体的代码实例：

【代码 1-5】（详见源代码目录 ch01-react-jsx-intro.html 文件）

```
01    <!DOCTYPE html>
02    <html>
03    <head>
04        <meta charset="UTF-8" />
05        <title>React JSX Intro</title>
06        <script src="https://unpkg.com/react@16/umd/react.development.js"></script>
07        <script src="https://unpkg.com/react-dom@16/umd/react-dom.development.js"></script>
08        <!-- Don't use this in production: -->
```

第 1 章 React 简介

```
09        <script src="https://unpkg.com/babel-standalone@6.15.0/babel.min.js"></script>
10    </head>
11    <body>
12        <!-- 添加文档主体内容 -->
13        <div id='id-div-react'></div>
14        <script type="text/babel">
15            // TODO: get div
16            var divReact = document.getElementById('id-div-react');
17            // TODO: React JSX
18            const reactSpan = (
19                <span>
20                    <h3>React JSX</h3>
21                    <p>Create React DOM by React JSX.</p>
22                </span>
23            );
24            ReactDOM.render(reactSpan, divReact);
25        </script>
26    </body>
27 </html>
```

关于【代码 1-5】的说明：

- 首先，在第 14 行代码中，<script type="text/babel">标签内使用了 JSX 语法要求的 "text/babel"属性，这一点会在后续章节中进行详细介绍。
- 第 18～23 行定义了一段完整的 JSX 代码，实现了一个虚拟 DOM 对象，具体内容如下：
  - 第 18～23 行代码通过 const 关键字定义了一个常量（reactSpan），该常量使用小括号包含了通过<span>、<h3>和<p>标签定义的元素组合。
  - 第 19～22 行代码定义的 HTML 标签符合 XML 格式，而常量（reactSpan）的定义完完全全符合 JavaScript 语法，因此该语法被称为 JSX。
- 第 24 行代码调用 React DOM 对象的 render()方法，将 JSX 代码渲染到页面中进行显示。

下面使用 Firefox 浏览器运行测试该 HTML 网页，具体效果如图 1.10 所示。

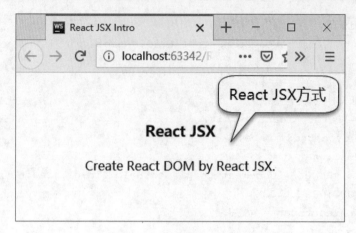

图 1.10　React JSX 简单入门

图 1.10 中显示了通过 React JSX 方式渲染的效果。其实，通过 React JSX 方式定义的虚拟 DOM，最终也会转换为通过 createElement()方法实现虚拟 DOM。

1.7　React 渲染机制

React 框架之所以大受欢迎，其特有的渲染机制是非常重要的因素之一。既然提到 React 渲染机制，那么就说一说 Diff 算法，该算法是支撑 React 渲染机制的核心技术之一。

Diff 算法博大精深、涉及的知识点很多，建议感兴趣的读者找专业的算法书籍作深入学习。这里，笔者就尝试用通俗的语言来描述一下，通过 Diff 算法实现 React 渲染机制的过程。

Diff 算法的核心就是通过比较找到 DOM Tree 前后的差异。React 渲染机制的基本原理就是：在 DOM Tree 的状态和属性发生改变后，构造出新的虚拟 DOM Tree，然后通过 Diff 算法与原始的虚拟 DOM Tree 进行比较，计算出变化的节点并进行更新操作。该算法的优势就是减少了对 DOM 的频繁重复操作，从而提升页面的访问性能。

为了验证 React 渲染机制是针对 DOM Tree 中的局部节点进行更新操作，我们看一个具体的代码实例：

【代码 1-6】（详见源代码目录 ch01-react-dom-render.html 文件）

```
01   <!DOCTYPE html>
02   <html>
03   <head>
04       <meta charset="UTF-8" />
05       <title>React render method</title>
06       <script src="https://unpkg.com/react@16/umd/react.development.js"></script>
07       <script src="https://unpkg.com/react-dom@16/umd/react-dom.development.js"></script>
```

```
08      <!-- Don't use this in production: -->
09      <script src="https://unpkg.com/babel-standalone@6.15.0/babel.min.js"></script>
10    </head>
11    <body>
12      <!-- 添加文档主体内容 -->
13      <div id="id-div-react"></div>
14      <script type="text/babel">
15        /**
16         * update time
17         */
18        function updateTime() {
19          const renderDiv= (<div>
20              <h3>React 渲染机制</h3>
21              <p>现在时间是 {new Date().toLocaleTimeString()}.</p>
22          </div>);
23          // TODO: get div
24          var divReact = document.getElementById('id-div-react');
25          // TODO: render div
26          ReactDOM.render(renderDiv, divReact);
27        }
28        // TODO: define timer
29        setInterval(updateTime, 1000);
30      </script>
31    </body>
32 </html>
```

关于【代码 1-6】的说明：

- 第 13 行代码通过<div id="id-div-react">标签元素定义了一个层，用于显示通过 React 框架渲染的文本内容。
- 第 18~27 行代码定义了一个自定义方法 updateTime()，用于实现通过 React 渲染更新元素，具体内容如下：
  - 第 19~22 行代码通过 const 关键字定义了一个常量（renderDiv），描述了要引入的容器节点<div>，包括一个<h3>标签元素用于定义标题和一个<p>标签元素定义的标题内容；同时，在<p>标签元素内使用花括号定义了一个时间对象，用于获取当前时间。
  - 第 24 行代码获取了页面中要渲染的元素节点<div id="id-div-react">，保存在变量（divReact）中。
  - 第 26 行代码调用 React DOM 对象的 render()方法，将虚拟 DOM（renderDiv）渲染到<div>层元素（divReact）中。

- 第 29 行代码使用 setInterval()方法设置了一个计时器，通过调用 updateTime()方法实现定时（1000ms）渲染更新元素。

下面使用 Firefox 浏览器运行测试该 HTML 网页，具体效果如图 1.11 所示。如图上的箭头和标识所示，页面效果显示通过 React 渲染方式，仅仅只更新时间元素节点。

图 1.11　React 实现渲染更新元素

第 2 章
◀ React JSX ▶

React JSX 是专门用于 React 框架的 JavaScript 语法扩展语言,其本质上也是一种 JavaScript 脚本语言。

2.1 JSX 介绍

JSX 其实就是 JavaScript XML 的缩写,直译过来就是基于 JavaScript 的 XML。同时,JSX 作为一种 JavaScript 语法扩展,支持自定义属性,并具有很强的扩展性。由于 JSX 是 React 框架内置的语法,且专用于 React 应用开发,因此建议设计人员使用 JSX 方式来实现 UI 中的虚拟 DOM。

若要在 React 项目中使用 JSX 语法,则必须引用"babel.js"来解析 JSX,且在<script>标签中必须改用"type="text/babel""属性。这里解释一下这么做的原因,在使用"type="text/babel""属性替换"type="text/javascript""属性后,浏览器内置的 JavaScript 解释器就不会解析<script>标签里的脚本代码,转而由"babel.js"进行解析,从而避免 React 代码与原生 JavaScript 代码发生混淆。

介绍了以上关于 JSX 的知识点,下面看一下 JSX 的一般语法形式。

```
const element= (<tag-level-1>
        <tag-level-2>
        ……
        </tag-level-2>
        </tag-level-1>);
```

这里使用 const 关键字定义常量名(element),表示 JSX 代码的名称。<tag-level-i>标签元素表示 DOM 标签,且支持多级嵌套的形式。

2.2 JSX 独立文件

JSX 可以在页面中通过直接定义(见【代码 1-5】)的方式实现,也可以通过引入外部定义好的 JSX 独立文件方式来实现。

下面尝试将【代码 1-5】以独立文件的方式进行改写，来实现 React JSX 代码的应用。

【代码 2-1】（详见源代码目录 import.js 文件）

```
01  // TODO: get div
02  var divReact = document.getElementById('id-div-react');
03  // TODO: React JSX
04  const reactSpan = (
05      <span>
06          <h3>React JSX</h3>
07          <p>Create React DOM by React JSX.</p>
08      </span>
09  );
10  // TODO: React render
11  ReactDOM.render(reactSpan, divReact);
```

关于【代码 2-1】的说明：

- 第 01～11 行定义了一段完整的 JSX 代码，实现了一个虚拟 DOM 对象，具体内容如下：
  - 第 02 行代码获取了层（<div id='id-div-react'>）标签元素的对象（divReact），该标签元素定义在下面的【代码 2-2】页面文件中。
  - 第 04～09 行代码通过 const 关键字定义了一个常量（reactSpan），该常量使用小括号包含了通过<span>、<h3>和<p>标签定义的元素组合，这段代码的定义方式就是 JSX 语法形式。
  - 第 11 行代码调用 React DOM 对象的 render()方法，将 JSX 代码渲染到页面中进行显示。

下面再看一下 HTML 页面文件的具体代码：

【代码 2-2】（详见源代码目录 ch02-react-jsx-import.html 文件）

```
01  <!DOCTYPE html>
02  <html>
03  <head>
04      <meta charset="UTF-8" />
05      <title>React JSX Intro</title>
06      <script src="https://unpkg.com/react@16/umd/react.development.js"></script>
07      <script src="https://unpkg.com/react-dom@16/umd/react-dom.development.js"></script>
08      <!-- Don't use this in production: -->
```

```
09      <script src="https://unpkg.com/babel-standalone@6.15.0/babel.min.js"></script>
10    </head>
11    <body>
12    <!-- 添加文档主体内容 -->
13    <div id='id-div-react'></div>
14    <script type="text/babel" src="import.js"></script>
15    </body>
16 </html>
```

关于【代码 2-2】的说明：

- 第 13 行代码中定义了一个层（<div id='id-div-react'>）标签元素。
- 在第 14 行代码中，通过<script>标签引入了 JSX 独立文件，其中 "src" 属性定义了 JSX 独立文件的路径，另外 "type" 属性定义为通过 Babel 方式（"text/babel"）进行解析。

下面使用 Firefox 浏览器运行测试该 HTML 网页，具体效果如图 2.1 所示。如图中的标识所示，页面中显示了通过 React JSX 独立文件方式渲染的效果，与【代码 1-5】实现的功能是一致的。

图 2.1　React JSX 独立文件方式

2.3　JSX 算术表达式

既然 React JSX 使用的就是 JavaScript 语法，那么自然也可以使用 JavaScript 表达式了。这里需要注意的是，在 React JSX 中使用 JavaScript 表达式要使用大括号 "｛｝" 括起来。

React JSX 中的 JavaScript 表达式有很多种形式，先看一个最简单的 JavaScript 算术表达式的代码实例，具体如下：

【代码 2-3】（详见源代码目录 ch02-react-jsx-exp-arithmetic.html 文件）

```
01 <!DOCTYPE html>
```

```
02  <html>
03  <head>
04      <meta charset="UTF-8"/>
05      <title>React JSX Exp - Arithmetic</title>
06      <script src="https://unpkg.com/react@16/umd/react.development.js"></script>
07      <script src="https://unpkg.com/react-dom@16/umd/react-dom.development.js"></script>
08      <!-- Don't use this in production: -->
09      <script src="https://unpkg.com/babel-standalone@6.15.0/babel.min.js"></script>
10  </head>
11  <body>
12  <!-- 添加文档主体内容 -->
13  <div id='id-div-react'></div>
14  <script type="text/babel">
15      // TODO: get div
16      var divReact = document.getElementById('id-div-react');
17      // TODO: React JSX
18      const reactSpan = (
19          <span>
20              <h3>JSX Expression - Arithmetic</h3>
21              <p>Now is calculating : 3 + 6 = {3 + 6}.</p>
22          </span>
23      );
24      // TODO: React render
25      ReactDOM.render(reactSpan, divReact);
26  </script>
27  </body>
28  </html>
```

关于【代码 2-3】的说明：

- 第 21 行代码中大括号"{3+6}"内定义的就是一个 JavaScript 计算表达式，React JSX 语法会将"3+6"的算术运算结果显示在页面中。

测试网页的效果如图 2.2 所示。页面中成功显示了表达式"3+6"的运算结果"9"。

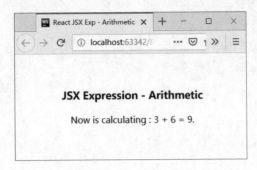

图 2.2 React JSX 中的 JavaScript 计算表达式

2.4 JSX 条件表达式

React JSX 中的 JavaScript 表达式是无法使用 if 条件语句的，但是可以使用三元条件表达式来替代 if 条件语句，具体代码如下：

【代码 2-4】（详见源代码目录 ch02-react-jsx-exp-tri-conditional.html 文件）

```
01  <!DOCTYPE html>
02  <html>
03  <head>
04      <meta charset="UTF-8"/>
05      <title>React JSX Exp - Triple Conditional</title>
06      <script src="https://unpkg.com/react@16/umd/react.development.js"></script>
07      <script src="https://unpkg.com/react-dom@16/umd/react-dom.development.js"></script>
08      <!-- Don't use this in production: -->
09      <script src="https://unpkg.com/babel-standalone@6.15.0/babel.min.js"></script>
10  </head>
11  <body>
12  <!-- 添加文档主体内容 -->
13  <div id='id-div-react'></div>
14  <script type="text/babel">
15      // TODO: get div
16      var divReact = document.getElementById('id-div-react');
17      // TODO: React JSX
18      const reactSpan = (
```

```
19            <span>
20                <h3>JSX Expression - Triple Conditional</h3>
21                <p>Test exp "1 == 1", return -> {1 == 1 ? "true" : "false"}.</p>
22                <p>Test exp "1 != 1", return -> {1 != 1 ? "true" : "false"}.</p>
23            </span>
24        );
25        // TODO: React render
26        ReactDOM.render(reactSpan, divReact);
27    </script>
28 </body>
29 </html>
```

关于【代码 2-4】的说明：

- 第 21 行代码中大括号"{ 1 == 1 ? 'true' : 'false' }"内定义的就是第一个 JavaScript 三元条件表达式。
- 第 22 行代码中大括号"{ 1 != 1 ? 'true' : 'false' }"内定义的就是第二个 JavaScript 三元条件表达式。

测试网页的效果如图 2.3 所示。如图中的箭头所示，测试"1 == 1"的三元条件表达式的运算结果为"true"，而测试"1 != 1"的三元条件表达式的运算结果为"false"。由此可见，JSX 是完全支持三元条件表达式运算的。

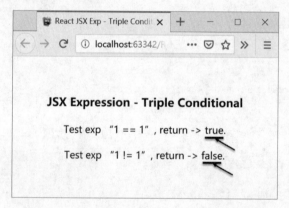

图 2.3　React JSX 中的 JavaScript 三元条件表达式

2.5　JSX 嵌入表达式

React JSX 可以支持嵌入表达式，这具体是个什么概念呢？简单来讲，如果我们先声明定义了一个变量，就可以通过在 JSX 中引用该变量的方式进行使用。下面看一个具体的代码实例：

【代码 2-5】（详见源代码目录 ch02-react-jsx-exp-embed.html 文件）

```html
01  <!DOCTYPE html>
02  <html>
03  <head>
04      <meta charset="UTF-8"/>
05      <title>React JSX Exp - Embed</title>
06      <script src="https://unpkg.com/react@16/umd/react.development.js"></script>
07      <script src="https://unpkg.com/react-dom@16/umd/react-dom.development.js"></script>
08      <!-- Don't use this in production: -->
09      <script src="https://unpkg.com/babel-standalone@6.15.0/babel.min.js"></script>
10  </head>
11  <body>
12  <!-- 添加文档主体内容 -->
13  <div id='id-div-react'></div>
14  <script type="text/babel">
15      // TODO: get div
16      var divReact = document.getElementById('id-div-react');
17      // TODO: define const
18      const name = "King";
19      // TODO: React JSX
20      const reactSpan = (
21          <span>
22              <h3>JSX Expression - Embed Expression</h3>
23              <p>Username : {name}.</p>
24          </span>
25      );
26      // TODO: React render
27      ReactDOM.render(reactSpan, divReact);
28  </script>
29  </body>
30  </html>
```

关于【代码 2-5】的说明：

- 第 18 行代码中通过 const 运算符定义了一个常量（name），并初始化为字符串"King"。
- 第 20～25 行代码定义了 JSX 常量（reactSpan），其中在第 23 行代码中直接通过表达式"{name}"方式嵌入了第 19 行代码定义的常量（name）。

测试网页的效果如图 2.4 所示。如图中的箭头所示，嵌入 JSX 的表达式"{name}"成功显示出了预先所定义的内容（King）。

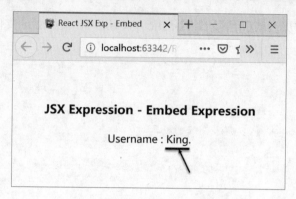

图 2.4　React JSX 嵌入表达式

2.6　JSX 对象表达式

在 React JSX 中可以直接使用对象表达式，也就是说完全支持 obj.property 的表达式形式。下面看一个具体的代码实例：

【代码 2-6】（详见源代码目录 ch02-react-jsx-exp-obj.html 文件）

```
01  <!DOCTYPE html>
02  <html>
03  <head>
04      <meta charset="UTF-8"/>
05      <title>React JSX Exp - Object</title>
06      <script src="https://unpkg.com/react@16/umd/react.development.js"></script>
07      <script src="https://unpkg.com/react-dom@16/umd/react-dom.development.js"></script>
08      <!-- Don't use this in production: -->
09      <script src="https://unpkg.com/babel-standalone@6.15.0/babel.min.js"></script>
10  </head>
11  <body>
12  <!-- 添加文档主体内容 -->
13  <div id='id-div-react'></div>
14  <script type="text/babel">
15      // TODO: get div
16      var divReact = document.getElementById('id-div-react');
17      // TODO: define object
18      const userinfo = {
19          name: "King",
```

```
20          gender: "male",
21          age: "18"
22      };
23      // TODO: React JSX
24      const reactSpan = (
25          <span>
26              <h3>JSX Expression - Object Expression</h3>
27      <p>Userinfo : name is {userinfo.name}, {userinfo.gender} and {userinfo.age} years old.</p>
28          </span>
29      );
30      // TODO: React render
31      ReactDOM.render(reactSpan, divReact);
32  </script>
33  </body>
34  </html>
```

关于【代码 2-6】的说明：

- 第 18～22 行代码中通过 const 运算符定义了一个常量对象（userinfo），并初始化了一组属性。
- 在第 27 行代码定义的 JSX 中，分别通过对象表达式"{userinfo.name}""{userinfo.gender}"和"{userinfo.age}"的方式使用了常量对象（userinfo）的属性值。

测试网页的效果如图 2.5 所示。如图中的箭头所示，在 JSX 中使用的一组对象表达式分别显示出了预先所定义的内容。

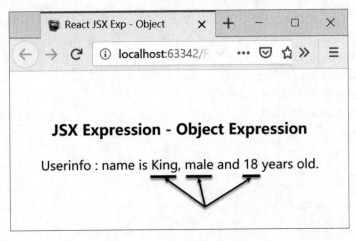

图 2.5　React JSX 对象表达式

2.7 JSX 函数表达式

在 React JSX 中还可以直接调用 JavaScript 函数，这就是 JSX 函数表达式。JSX 函数表达式应用起来非常简单、灵活，具有很强大的功能。为了更好地进行对比，我们将【代码 2-6】按照函数表达式的形式进行改写，具体代码实例如下：

【代码 2-7】（详见源代码目录 ch02-react-jsx-exp-func.html 文件）

```
01  <!DOCTYPE html>
02  <html>
03  <head>
04      <meta charset="UTF-8"/>
05      <title>React JSX Exp - Functional</title>
06      <script src="https://unpkg.com/react@16/umd/react.development.js"></script>
07      <script src="https://unpkg.com/react-dom@16/umd/react-dom.development.js"></script>
08      <!-- Don't use this in production: -->
09      <script src="https://unpkg.com/babel-standalone@6.15.0/babel.min.js"></script>
10  </head>
11  <body>
12  <!-- 添加文档主体内容 -->
13  <div id='id-div-react'></div>
14  <script type="text/babel">
15      // TODO: get div
16      var divReact = document.getElementById('id-div-react');
17      // TODO: define object
18      const userinfo = {
19          name: "King",
20          gender: "male",
21          age: "18"
22      };
23      // TODO: function
24      function formatUserinfo(ui) {
25      return "Userinfo : name is " + ui.name + ", " + ui.gender + " and " + ui.age + " years old.";
26      }
27      // TODO: React JSX
28      const reactSpan = (
29          <span>
30              <h3>JSX Expression - Functional Expression</h3>
```

```
31            <p>{formatUserinfo(userinfo)}</p>
32        </span>
33    );
34    // TODO: React render
35    ReactDOM.render(reactSpan, divReact);
36 </script>
37 </body>
38 </html>
```

关于【代码 2-7】的说明：

- 第 18~22 行代码中通过 const 运算符定义了一个常量对象（userinfo），并初始化了一组属性。
- 第 24~26 行代码定义了一个函数（formatUserinfo），该函数返回了一个将常量对象（userinfo）进行组合后的字符串信息。
- 在第 31 行代码定义的 JSX 中，通过函数表达式 "formatUserinfo(userinfo)" 的方式获取了常量对象（userinfo）的信息。

测试网页的效果如图 2.6 所示。如图中的箭头所示，JSX 函数表达式成功显示出了常量对象（userinfo）中定义的内容。

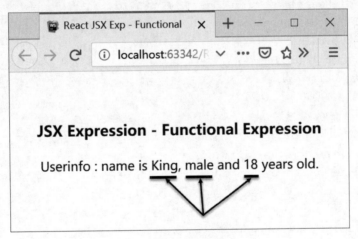

图 2.6　React JSX 函数表达式

2.8　JSX 增强函数表达式

在前文中我们知道，在 React JSX 中是无法直接使用 if 条件语句的。不过，可以通过 JSX 函数表达式以变通的方式来使用。因为，JSX 可以作为参数传入并嵌套在另一个 JSX 之中。为了更好地进行对比，我们再将【代码 2-7】进一步做改写，具体代码实例如下：

【代码 2-8】（详见源代码目录 ch02-react-jsx-exp-func-plus.html 文件）

```html
01   <!DOCTYPE html>
02   <html>
03   <head>
04       <meta charset="UTF-8"/>
05       <title>React JSX Exp - Functional</title>
06       <script src="https://unpkg.com/react@16/umd/react.development.js"></script>
07       <script src="https://unpkg.com/react-dom@16/umd/react-dom.development.js"></script>
08       <!-- Don't use this in production: -->
09       <script src="https://unpkg.com/babel-standalone@6.15.0/babel.min.js"></script>
10   </head>
11   <body>
12   <!-- 添加文档主体内容 -->
13   <div id='id-div-react'></div>
14   <script type="text/babel">
15       // TODO: get div
16       var divReact = document.getElementById('id-div-react');
17       // TODO: define object
18       const userinfo = {
19           name: "King",
20           gender: "male",
21           age: "18"
22       };
23       // TODO: function - formatUserinfo
24       function formatUserinfo(ui) {
25       return "Userinfo : name is " + ui.name + ", " + ui.gender + " and " + ui.age + " years old.";
26       }
27       // TODO: function - formatUserinfo
28       function chooseUserinfo(ui) {
29           if(ui) {
30       return "Userinfo : name is " + ui.name + ", " + ui.gender + " and " + ui.age + " years old.";
31           } else {
32               return "Hello, userinfo is nothing.";
33           }
34       }
35       // TODO: React JSX
36       const reactSpan = (
37           <span>
```

```
38              <h3>JSX Expression - Plus Functional Expression</h3>
39              <p>{chooseUserinfo(userinfo)}</p>
40              <p>{chooseUserinfo()}</p>
41          </span>
42      );
43      // TODO: React render
44      ReactDOM.render(reactSpan, divReact);
45  </script>
46  </body>
47  </html>
```

关于【代码 2-8】的说明：

- 第 18~22 行代码通过 const 运算符定义了一个常量对象（userinfo），并初始化了一组属性。
- 第 24~26 行代码定义了第一个函数（formatUserinfo），该函数返回了一个将常量对象（userinfo）进行组合后的字符串信息。
- 第 28~34 行代码定义了第二个函数（chooseUserinfo），该函数通过 if 条件语句判断传入的参数（ui）。若判断结果为"true"，则调用 formatUserinfo()方法返回一个将常量对象（userinfo）进行组合后的字符串信息。若判断结果为"false"，则直接返回"Hello, userinfo is nothing（用户信息为空）"的提示信息。
- 在第 39 行代码定义的 JSX 中，通过带参数的函数表达式（chooseUserinfo(userinfo)）方式获取了用户信息。
- 在第 40 行代码定义的 JSX 中，通过无参数的函数表达式（chooseUserinfo()）方式获取了相应的信息。

测试网页的效果如图 2.7 所示。如图中的箭头所示，通过无参数的函数表达式（chooseUserinfo()）方式获取了"Hello, userinfo is nothing（用户信息为空）"的提示信息。

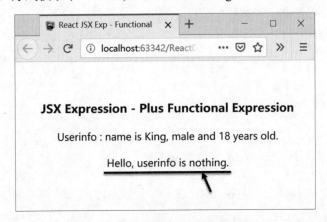

图 2.7　React JSX 增强函数表达式

2.9 JSX 数组表达式

在 React JSX 中除了可以直接使用对象表达式，还可以使用数组表达式的形式。为了更好地进行对比，我们将【代码 2-5】进行了改写，具体代码实例如下：

【代码 2-9】（详见源代码目录 ch02-react-jsx-exp-arr.html 文件）

```
01  <!DOCTYPE html>
02  <html>
03  <head>
04      <meta charset="UTF-8"/>
05      <title>React JSX Exp - Array</title>
06      <script src="https://unpkg.com/react@16/umd/react.development.js"></script>
07      <script src="https://unpkg.com/react-dom@16/umd/react-dom.development.js"></script>
08      <!-- Don't use this in production: -->
09      <script src="https://unpkg.com/babel-standalone@6.15.0/babel.min.js"></script>
10  </head>
11  <body>
12  <!-- 添加文档主体内容 -->
13  <div id='id-div-react'></div>
14  <script type="text/babel">
15      // TODO: get div
16      var divReact = document.getElementById('id-div-react');
17      // TODO: define array
18      var arrUserinfo = [
19          <span>Name id king,</span>,
20          <span>male,</span>,
21          <span>and 18 years old.</span>
22      ];
23      // TODO: React JSX
24      const reactSpan = (
25          <span>
26              <h3>JSX Expression - Array Expression</h3>
27              <p>{arrUserinfo}</p>
28          </span>
29      );
30      // TODO: React render
31      ReactDOM.render(reactSpan, divReact);
32  </script>
```

```
33      </body>
34  </html>
```

关于【代码2-9】的说明：

- 第 18~22 行代码通过 var 运算符定义了一个数组（arrUserinfo），并初始化了一组 \<span\>标签元素。
- 在第 27 行代码定义的 JSX 中，直接通过数组表达式 "{arrUserinfo}" 的方式获取了数组（arrUserinfo）中的数据，JSX 语法会自动将全部数组项展开。

测试网页的效果如图 2.8 所示。

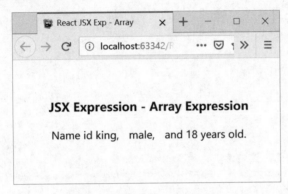

图 2.8　React JSX 数组表达式

2.10　JSX 样式表达式

React 语法支持内联形式的 CSS 样式，因此可以使用 JSX 样式表达式来实现页面风格。下面看一个具体的代码实例：

【代码2-10】（详见源代码目录 ch02-react-jsx-exp-style.html 文件）

```
01  <!DOCTYPE html>
02  <html>
03  <head>
04      <meta charset="UTF-8"/>
05      <title>React JSX Exp - Style</title>
06      <script src="https://unpkg.com/react@16/umd/react.development.js"></script>
07      <script src="https://unpkg.com/react-dom@16/umd/react-dom.development.js"></script>
08      <!-- Don't use this in production: -->
09      <script src="https://unpkg.com/babel-standalone@6.15.0/babel.min.js"></script>
```

```html
10  </head>
11  <body>
12  <!-- 添加文档主体内容 -->
13  <div id='id-div-react'></div>
14  <script type="text/babel">
15      // TODO: get div
16      var divReact = document.getElementById('id-div-react');
17      // TODO: define css
18      const css_p_lg = {
19          fontSize: 20,
20          fontStyle: "bold",
21          color: "red"
22      };
23      const css_p_md = {
24          fontSize: 16,
25          fontStyle: "normal",
26          color: "green"
27      };
28      const css_p_sm = {
29          fontSize: 12,
30          fontStyle: "italic",
31          color: "blue"
32      };
33      // TODO: React JSX
34      const reactSpan = (
35          <span>
36              <h3>JSX Expression - Style Expression</h3>
37              <p style={css_p_lg}>JSX Style Expression</p>
38              <p style={css_p_md}>JSX Style Expression</p>
39              <p style={css_p_sm}>JSX Style Expression</p>
40          </span>
41      );
42      // TODO: React render
43      ReactDOM.render(reactSpan, divReact);
44  </script>
45  </body>
46  </html>
```

关于【代码 2-10】的说明：

- 第 18~22 行代码、第 23~27 行代码和第 28~32 行代码，分别通过 const 运算符定义了三组对象（css_p_lg、css_p_md 和 css_p_sm），每个对象均定义了不同的 CSS 样式。

- 在第 37~39 行定义的 JSX 代码中，实现了一组段落<p>标签元素。在每个<p>标签元素内，通过 style 属性使用了 JSX 样式表达式，将前面定义的样式对象（css_p_lg、css_p_md 和 css_p_sm）应用在这一组段落内容中。

测试网页的效果如图 2.9 所示。如图中的标识所示，段落<p>标签元素中的内容通过 JSX 样式表达式实现了不同的样式风格。

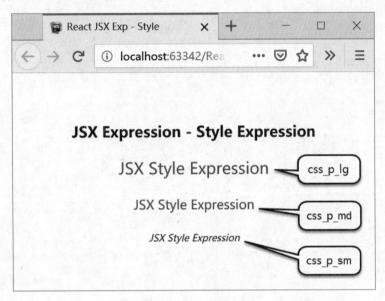

图 2.9　React JSX 样式表达式

2.11　JSX 注释表达式

如何在 React JSX 中使用注释呢？这一点 React 语法的定义有点特别，注释内容也需要放在大括号 "{ }" 之中使用，否则注释的内容就会也页面内容的形式显示出来了。下面看一个具体的代码实例：

【代码 2-11】（详见源代码目录 ch02-react-jsx-exp-comment.html 文件）

```
01  <!DOCTYPE html>
02  <html>
03  <head>
04      <meta charset="UTF-8"/>
05      <title>React JSX Exp - Comment</title>
06      <script src="https://unpkg.com/react@16/umd/react.development.js"></script>
07      <script src="https://unpkg.com/react-dom@16/umd/react-dom.development.js"></script>
```

```
08          <!-- Don't use this in production: -->
09          <script src="https://unpkg.com/babel-standalone@6.15.0/babel.min.js"></script>
10      </head>
11      <body>
12          <!-- 添加文档主体内容 -->
13          <div id='id-div-react'></div>
14          <script type="text/babel">
15              // TODO: get div
16              var divReact = document.getElementById('id-div-react');
17              // TODO: React JSX
18              const reactSpan = (
19                  <span>
20                      <h3>JSX Expression - Comment Expression</h3>
21                      {/* This is a right JSX Comment Expression. */}
22                      /* This is a wrong JSX Comment Expression. */
23                  </span>
24              );
25              // TODO: React render
26              ReactDOM.render(reactSpan, divReact);
27          </script>
28      </body>
29  </html>
```

关于【代码 2-11】的说明：

- 在第 18~24 行定义的 JSX 代码中，第 21 行代码中放进大括号 "{ }" 中的注释内容是正确的形式，而第 22 行代码中没使用大括号 "{ }" 的注释内容是不正确的形式，将会以页面内容的形式进行显示。

测试网页的效果如图 2.10 所示。如图中的箭头所示，第 22 行代码中没使用大括号 "{ }" 的注释内容在页面中显示出来了，这自然不是设计人员想要看到的效果。

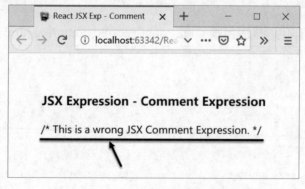

图 2.10　React JSX 注释表达式

第 3 章

◀React组件与Props▶

　　React 组件可以将 UI 切分成一些独立的、可复用的部件,这样有助于设计人员专注于构建每一个单独的部件。React 组件通过 Props 可以接收任意的输入值,因此 Props 也可以理解为参数的概念。

3.1　React 组件介绍

　　React 组件从形式上看很像是 JavaScript 函数,通过这个函数返回一个需要在页面上展示的 React 元素。其实,在第 2 章的代码实例中我们已经使用到了 React 组件,具体的代码如下:

【代码 3-1】

```
function reactComponent() {
  return <tagName>Hello, React Component.</tagName>;
}
```

　　上面这段代码就是通过 JavaScript 函数形式实现的 React 组件。React 组件的结构看起来十分简洁和高效,在函数内部定义好页面需要的元素组合,通过 return 语句返回函数值就可以了。

　　另外,在本书第 1 章中介绍过,React 语法是基于版本 ECMAScript 6 实现的。因此,React 组件除了通过 JavaScript 函数的形式,还可以通过 ES6 Class(类)的形式来实现,具体代码如下:

【代码 3-2】

```
class reactComponent extends React.Component {
  render() {
    return <tagName>Hello, React Component.</tagName>;
  }
}
```

　　上面这段代码就是通过 ES6 Class(类)形式实现的 React 组件。其实,【代码 3-2】与【代码 3-1】虽然实现的手段不一样,但最终的效果是一致的。

3.2 React 函数组件

本节先介绍如何通过 JavaScript 函数形式实现一个 React 函数组件。下面看一个具体的代码实例：

【代码 3-3】（详见源代码目录 ch03-react-comp-func.html 文件）

```
01  <!DOCTYPE html>
02  <html>
03  <head>
04      <meta charset="UTF-8"/>
05      <title>React Component - Function</title>
06      <script src="https://unpkg.com/react@16/umd/react.development.js"></script>
07      <script src="https://unpkg.com/react-dom@16/umd/react-dom.development.js"></script>
08      <!-- Don't use this in production: -->
09      <script src="https://unpkg.com/babel-standalone@6.15.0/babel.min.js"></script>
10  </head>
11  <body>
12  <!-- 添加文档主体内容 -->
13  <div id='id-div-react'></div>
14  <script type="text/babel">
15      // TODO: get div
16      var divReact = document.getElementById('id-div-react');
17      // TODO: function component
18      function HelloReactComp() {
19          return <p>Hello, this is a React functional component.</p>;
20      }
21      // TODO: define const
22      const eleHello = <HelloReactComp/>;
23      // TODO: React JSX
24      const reactSpan = (
25          <span>
26              <h3>React Component - Function Component</h3>
27              {eleHello}
28          </span>
29      );
30      // TODO: React render
31      ReactDOM.render(reactSpan, divReact);
32  </script>
```

```
33    </body>
34    </html>
```

关于【代码 3-3】的说明：

- 第 18~20 行代码定义了一个 JavaScript 函数（HelloReactComp()），该函数就是我们要实现的 React 函数组件，具体说明如下：
  - 第 19 行代码通过 return 语句返回了一行 JSX 代码，该行代码通过<p>标签元素定义了一行段落文本。
- 关键看第 22 行代码，通过 const 关键字定义了一个常量（eleHello），然后将 JavaScript 函数（HelloReactComp()）的名称（HelloReactComp）使用尖括号"<>"包括进去，赋值给常量（eleHello），相当于将函数名称（HelloReactComp）作为标签名来使用。通过以上步骤的定义，JavaScript 函数（HelloReactComp()）就成为了一个 React 函数组件（<HelloReactComp>）。
- 第 27 行代码通过常量（eleHello）放入 JSX 代码（{eleHello}）来实现 React 函数组件的应用。

测试网页的效果如图 3.1 所示。

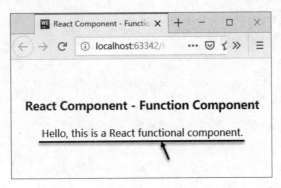

图 3.1　React 函数组件

如图 3.1 中箭头所示，页面中显示了通过 React 函数组件渲染的效果，从效果来看与 JSX 代码一致。

尽管 React 组件与 JSX 代码实现的功能基本一致，但从设计角度上讲还是推荐使用 React 组件方式。原因是：将 React 组件与 Props 结合使用可以实现更灵活的功能，关于 Props 的内容会在本章后续内容中详细介绍。

3.3　React 类组件

本节接着介绍如何通过 ES6 Class（类）形式实现一个 React 类组件。下面看一个具体的代码实例：

【代码 3-4】（详见源代码目录 ch03-react-comp-class.html 文件）

```html
01  <!DOCTYPE html>
02  <html>
03  <head>
04      <meta charset="UTF-8"/>
05      <title>React Component - Class</title>
06      <script src="https://unpkg.com/react@16/umd/react.development.js"></script>
07      <script src="https://unpkg.com/react-dom@16/umd/react-dom.development.js"></script>
08      <!-- Don't use this in production: -->
09      <script src="https://unpkg.com/babel-standalone@6.15.0/babel.min.js"></script>
10  </head>
11  <body>
12  <!-- 添加文档主体内容 -->
13  <div id='id-div-react'></div>
14  <script type="text/babel">
15      // TODO: get div
16      var divReact = document.getElementById('id-div-react');
17      // TODO: class component
18      class HelloReactComp extends React.Component {
19          render() {
20              return <p>Hello, this is a React class component.</p>;
21          }
22      }
23      // TODO: define const
24      const eleHello = <HelloReactComp/>;
25      // TODO: React JSX
26      const reactSpan = (
27          <span>
28              <h3>React Component - Class Component</h3>
29              {eleHello}
30          </span>
31      );
32      // TODO: React render
33      ReactDOM.render(reactSpan, divReact);
```

```
34     </script>
35   </body>
36 </html>
```

关于【代码 3-4】的说明：

- 第 18~22 行代码定义了一个 ES6 Class 类（HelloReactComp），这个类就是我们要实现的 React 类组件，具体说明如下：
  - 第 18 行代码通过 class 关键字定义了类名（HelloReactComp），并通过 extends 关键字声明该类继承自 React.Component 对象。
  - 第 19~21 行代码通过 React.Component 对象的 render()方法实现 React 类组件的渲染操作，具体渲染内容是第 20 行代码返回的 JSX 代码。
- 第 24 行代码中，通过 const 关键字定义了一个常量（eleHello），然后将 ES6 Class 类（HelloReactComp）的类名（HelloReactComp）使用尖括号 "<>" 包括起来，赋值给常量（eleHello），类似于将类名（HelloReactComp）作为标签名来使用。通过以上步骤的定义，ES6 Class 类（HelloReactComp()）就成为 React 类组件（<HelloReactComp>）。
- 第 29 行代码中，通过将常量（eleHello）放入 JSX 代码（{eleHello}）来实现 React 类组件的应用。

测试网页的效果如图 3.2 所示。

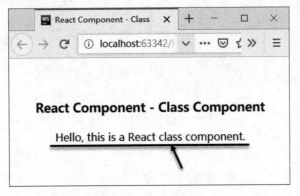

图 3.2　React 类组件

如图 3.2 中的箭头所示，页面中显示了通过 React 类组件渲染的效果，从效果来看与 React 函数组件功能相同。

另外注意一点，React 组件的名称首字母必须是大写的。例如，<HelloReactComp>是正确的，而<helloReactComp>就是非法的。这个规定主要是为了与原生 HTML 标签名称相区别，因为原生 HTML 标签名称均是小写字母开头的。

3.4 React 组合组件

React 组件支持在自身定义中引用其他组件，从而构成 React 组合组件（也称复合组件）。使用 React 组合组件定义的好处就是，可以使用同一组件来抽象出任意层次的细节，比如一个 React 表单组件就可以由多个不同类型的子组件来构成，还有对于对话框组件来说也是如此。

下面看一个通过 React 组合组件实现用户登录表单的代码实例，具体如下：

【代码 3-5】（详见源代码目录 ch03-react-comp-composing.html 文件）

```
01  <!DOCTYPE html>
02  <html>
03  <head>
04      <meta charset="UTF-8"/>
05      <title>React Component - Composing</title>
06      <script src="https://unpkg.com/react@16/umd/react.development.js"></script>
07      <script src="https://unpkg.com/react-dom@16/umd/react-dom.development.js"></script>
08      <!-- Don't use this in production: -->
09      <script src="https://unpkg.com/babel-standalone@6.15.0/babel.min.js"></script>
10  </head>
11  <body>
12  <!-- 添加文档主体内容 -->
13  <div id='id-div-react'></div>
14  <script type="text/babel">
15      // TODO: get div
16      var divReact = document.getElementById('id-div-react');
17      // TODO: function component
18      function FormTitle() {
19          return <h4>User Login</h4>;
20      }
21      function UserId() {
22          const userId = (
23              <p>User Id(*): <input type="text"/></p>
24          );
25          return userId;
26      }
27      function UserName() {
28          const userName = (
29              <p>User Name: <input type="text"/></p>
30          );
```

```
31            return userName;
32        }
33        function Password() {
34            const passwd = (
35                <p>Password(*): <input type="text"/></p>
36            );
37            return passwd;
38        }
39        function Submit() {
40            const submit = (
41                <p><button>Login</button></p>
42            );
43            return submit;
44        }
45        // TODO: composing component
46        function FormLogin() {
47            return (
48                <div id="id-div-formlogin">
49                    <FormTitle/>
50                    <UserId/>
51                    <UserName/>
52                    <Password/>
53                    <Submit/>
54                </div>
55            );
56        }
57        // TODO: define const
58        const formLogin = <FormLogin/>;
59        // TODO: React JSX
60        const reactSpan = (
61            <span>
62                <h3>React Component - Composing Component</h3>
63                {formLogin}
64            </span>
65        );
66        // TODO: React render
67        ReactDOM.render(reactSpan, divReact);
68    </script>
69    </body>
70    </html>
```

关于【代码 3-5】的说明：

- 第 18~44 行代码分别定义了若干 React 函数组件，用于构建一个 React 组合组件，该组合组件定义为一个用户登录表单，具体说明如下：

> 第 18~20 行代码定义的 React 函数组件（FormTitle()），用于定义表单名称。
> 第 21~26 行代码定义的 React 函数组件（UserId()），用于定义表单中的用户 id 域控件。
> 第 27~32 行代码定义的 React 函数组件（UserName()），用于定义表单中的用户名域控件。
> 第 33~38 行代码定义的 React 函数组件（Password()），用于定义表单中的密码域控件。
> 第 39~44 行代码定义的 React 函数组件（Submit()），用于定义表单中的提交按钮域控件。

- 关键是第 46~56 行代码定义的 React 组合组件（FormLogin()），其中第 49~53 行代码分别引用了第 18~38 行代码定义的若干 React 函数组件，用来组成表单中的各个用户域。
- 第 58 行代码中，通过 const 关键字定义了一个常量（formLogin），赋值为 React 组合组件（<FormLogin>）。
- 第 63 行代码中，通过将常量（formLogin）放入 JSX 代码（{formLogin}）来实现 React 组合组件的应用。

测试网页的效果如图 3.3 所示。如图中的标识所示，页面中显示了通过 React 组合组件实现的用户登录表单页面。

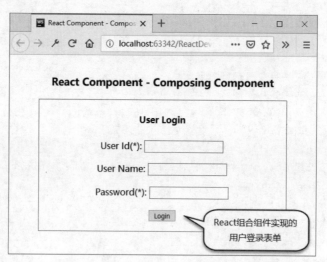

图 3.3　React 组合组件

3.5 React Props 介绍

既然 React 函数组件可以通过 JavaScript 函数方式实现,那么 React 函数组件就可以接受参数的传入。实际也确实如此,React 框架定义了一个 Props 的概念,专门用来实现 React 函数组件接受参数的输入。

下面看一个使用 React Props 实现参数输入的具体代码实例:

【代码 3-6】(详见源代码目录 ch03-react-props-intro.html 文件)

```
01  <!DOCTYPE html>
02  <html>
03  <head>
04      <meta charset="UTF-8"/>
05      <title>React Props - Intro</title>
06      <script src="https://unpkg.com/react@16/umd/react.development.js"></script>
07      <script src="https://unpkg.com/react-dom@16/umd/react-dom.development.js"></script>
08      <!-- Don't use this in production: -->
09      <script src="https://unpkg.com/babel-standalone@6.15.0/babel.min.js"></script>
10  </head>
11  <body>
12  <!-- 添加文档主体内容 -->
13  <div id='id-div-react'></div>
14  <script type="text/babel">
15      // TODO: get div
16      var divReact = document.getElementById('id-div-react');
17      // TODO: function component
18      function PropsReactComp(props) {
19          if(props) {
20              return <p>Hello, this is a React Props usage by {props.name}.</p>;
21          } else {
22              return <p>Hello, this is a React Props introduction.</p>;
23          }
24      }
25      // TODO: define const
26      const eleProps = <PropsReactComp name="King"/>;
27      // TODO: React JSX
28      const reactSpan = (
29          <span>
30              <h3>React Props - Introduction</h3>
```

```
31                  {eleProps}
32              </span>
33          );
34      // TODO: React render
35      ReactDOM.render(reactSpan, divReact);
36  </script>
37  </body>
38  </html>
```

关于【代码 3-6】的说明：

- 第 18～24 行代码分别定义了一个 React 函数组件（PropsReactComp()），注意在函数内定义了一个参数（props），具体说明如下：
  > 第 19～23 行代码通过条件语句判断参数（props）是否为"true"，若为"true"，则执行第 20 行代码。
  > 在第 20 行代码中，通过参数（props）对象获取其"name"属性值，这个"name"属性的定义在下面的第 26 行代码中。
- 第 26 行代码中，通过 const 关键字定义了一个常量（eleProps），赋值为 React 函数组件（<PropsReactComp>）。这里的特别之处在于，在 React 函数组件（<PropsReactComp>）中增加定义了一个"name"属性，同时初始化了属性值（"King"）。在前面的第 20 行代码中，就是通过参数（props）对象获取了该"name"属性的值。

测试网页的效果如图 3.4 所示。如图中的箭头所示，页面中显示了 props 参数对象获取了的"name"属性的值（King），这个就是 React Props 的基本使用方法。

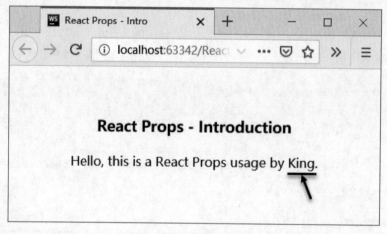

图 3.4 React Props 初步

3.6 React Props 应用

既然 React Props 可以接受参数的传入,它就可以在具体项目应用发挥很大的作用。下面看一个使用 React Props 参数实现表单初始化的代码实例,具体如下:

【代码 3-7】(详见源代码目录 ch03-react-props-form.html 文件)

```
01  <!DOCTYPE html>
02  <html>
03  <head>
04      <meta charset="UTF-8"/>
05      <title>React Props - Form</title>
06      <script src="https://unpkg.com/react@16/umd/react.development.js"></script>
07      <script src="https://unpkg.com/react-dom@16/umd/react-dom.development.js"></script>
08      <!-- Don't use this in production: -->
09      <script src="https://unpkg.com/babel-standalone@6.15.0/babel.min.js"></script>
10  </head>
11  <body>
12  <!-- 添加文档主体内容 -->
13  <div id='id-div-react'></div>
14  <script type="text/babel">
15      // TODO: get div
16      var divReact = document.getElementById('id-div-react');
17      // TODO: function component
18      function FormTitle() {
19          return <h4>User Login</h4>;
20      }
21      function UserId(props) {
22          const userId = (
23              <p>User Id(*): <input type="text" value={props.userId} readOnly/></p>
24          );
25          return userId;
26      }
27      function UserName(props) {
28          const userName = (
29              <p>User Name: <input type="text" value={props.userName} readOnly/></p>
30          );
```

```
31          return userName;
32      }
33      function Password() {
34          const passwd = (
35              <p>Password(*): <input type="password" /></p>
36          );
37          return passwd;
38      }
39      function Submit() {
40          const submit = (
41              <p><button>Login</button></p>
42          );
43          return submit;
44      }
45      // TODO: composing component
46      function FormLogin() {
47          return (
48              <div id="id-div-formlogin">
49                  <FormTitle />
50                  <UserId userId="007" />
51                  <UserName userName="King" />
52                  <Password />
53                  <Submit />
54              </div>
55          );
56      }
57      // TODO: define const
58      const formLogin = <FormLogin />;
59      // TODO: React JSX
60      const reactSpan = (
61        <span>
62            <h3>React Props - Form Init</h3>
63            {formLogin}
64        </span>
65      );
66      // TODO: React render
67      ReactDOM.render(reactSpan, divReact);
68  </script>
69  </body>
70  </html>
```

关于【代码 3-7】的说明：

- 这段代码是在【代码 3-5】的基础上修改而成的，主要是通过 React Props 参数传入实现表单的初始化操作。
- 第 18~44 行代码分别定义了若干 React 函数组件，用于构建一个 React 组合组件，该组合组件定义为一个用户登录表单。此处的定义与【代码 3-5】类似，不同之处说明如下：
  ➢ 在第 21~26 行代码定义的 React 函数组件（UserId()）中，增加了 Props 参数的定义。
  ➢ 在第 27~32 行代码定义的 React 函数组件（UserName()）中，也增加了 Props 参数的定义。
- 在第 46~56 行代码定义的 React 组合组件（FormLogin()）中，第 50 行和第 51 行代码分别增加了 "userId" 和 "userName" 属性的定义，同时初始化了属性值。

测试网页的效果如图 3.5 所示。如图中的箭头所示，页面表单中的 "User Id" 项和 "User Name" 项分别进行了初始化操作，分别获取了属性值（007）和属性值（King）。

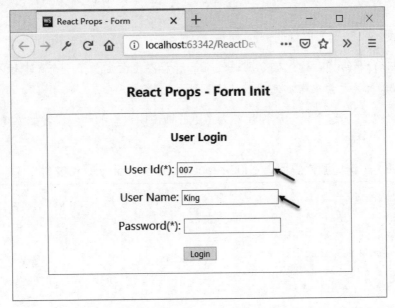

图 3.5　React Props 初始化表单项

3.7　React Props 只读性

虽然 React Props 很好用，但在使用上是有限制的。React 框架规定 Props 是不能被修改的，也就是说 Props 是个只读的参数。如果尝试修改 Props 会发生什么呢？下面，我们再回过头看一下【代码 3-7】定义的代码实例：

【代码 3-7 节选】（详见源代码目录 ch03-react-props-form.html 文件）

```
21    function UserId(props) {
22        const userId = (
23            <p>User Id(*): <input type="text" value={props.userId} readOnly/></p>
24        );
25        return userId;
26    }
27    function UserName(props) {
28        const userName = (
29            <p>User Name: <input type="text" value={props.userName} readOnly/></p>
30        );
31        return userName;
32    }
```

关于【代码 3-7 节选】的说明：

- 第 23 行和第 29 行代码分别通过<input>标签元素定义了一个文本输入框，注意在<input>标签元素中增加定义了只读属性（readOnly）。

这里为什么要增加这个只读属性呢？为了更好地说明这里的代码，我们将【代码 3-7】改写如下：

【代码 3-8】（详见源代码目录 ch03-react-props-readonly.html 文件）

```
01  <!DOCTYPE html>
02  <html>
03  <head>
04      <meta charset="UTF-8"/>
05      <title>React Props - Readonly</title>
06      <script src="https://unpkg.com/react@16/umd/react.development.js"></script>
07      <script src="https://unpkg.com/react-dom@16/umd/react-dom.development.js"></script>
08      <!-- Don't use this in production: -->
09      <script src="https://unpkg.com/babel-standalone@6.15.0/babel.min.js"></script>
10  </head>
11  <body>
12      <!-- 添加文档主体内容 -->
13      <div id='id-div-react'></div>
14      <script type="text/babel">
15          // TODO: get div
```

```
16      var divReact = document.getElementById('id-div-react');
17      // TODO: function component
18      function FormTitle() {
19          return <h4>User Login</h4>;
20      }
21      function UserId(props) {
22          const userId = (
23              <p>User Id(*): <input type="text" value={props.userId} /></p>
24          );
25          return userId;
26      }
27      function UserName(props) {
28          const userName = (
29              <p>User Name: <input type="text" value={props.userName} /></p>
30          );
31          return userName;
32      }
33      function Password() {
34          const passwd = (
35              <p>Password(*): <input type="password" /></p>
36          );
37          return passwd;
38      }
39      function Submit() {
40          const submit = (
41              <p><button>Login</button></p>
42          );
43          return submit;
44      }
45      // TODO: composing component
46      function FormLogin() {
47          return (
48              <div id="id-div-formlogin">
49                  <FormTitle />
50                  <UserId userId="007" />
51                  <UserName userName="King" />
52                  <Password />
53                  <Submit />
54              </div>
55          );
56      }
57      // TODO: define const
58      const formLogin = <FormLogin />;
```

```
59        // TODO: React JSX
60        const reactSpan = (
61          <span>
62            <h3>React Props - Readonly</h3>
63            {formLogin}
64          </span>
65        );
66        // TODO: React render
67        ReactDOM.render(reactSpan, divReact);
68      </script>
69    </body>
70  </html>
```

关于【代码 3-8】的说明：

- 在第 23 行和第 29 行代码通过<input>标签元素定义的文本输入框中，取消了只读属性（readOnly）的定义。

测试网页的效果如图 3.6 所示。如图中的标识所示，浏览器控制台中显示了关于 props 参数使用的错误信息。这里就不逐句翻译内容了，大致意思就是"Props 为只读类型，将其放入文本输入框这种可改变内容的域中是错误的做法"。

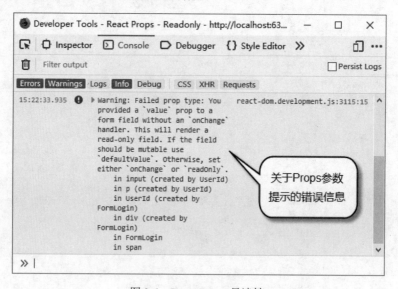

图 3.6　React Props 只读性

3.8　React Props 默认值

前面的 React Props 均是应用在 React 函数组件中的，这个自然也很好理解，因为定义

JavaScript 函数时是支持带参数的。那么问题就来了，在使用 React 类组件时如何使用 React Props 参数呢？

React 框架为类组件定义了一个默认 Props——defaultProps，使用 defaultProps 默认值来实现 React Props 应用。下面看一个 defaultProps 默认值的代码实例：

【代码 3-9】（详见源代码目录 ch03-react-props-defaultProps.html 文件）

```html
01  <!DOCTYPE html>
02  <html>
03  <head>
04      <meta charset="UTF-8"/>
05      <title>React Props - Default Props</title>
06      <script src="https://unpkg.com/react@16/umd/react.development.js"></script>
07      <script src="https://unpkg.com/react-dom@16/umd/react-dom.development.js"></script>
08      <!-- Don't use this in production: -->
09      <script src="https://unpkg.com/babel-standalone@6.15.0/babel.min.js"></script>
10  </head>
11  <body>
12  <!-- 添加文档主体内容 -->
13  <div id='id-div-react'></div>
14  <script type="text/babel">
15      // TODO: get div
16      var divReact = document.getElementById('id-div-react');
17      // TODO: function component
18      class PropsReactComp extends React.Component {
19          render() {
20              return <p>Hello, this is a React Props usage by {this.props.default}.</p>;
21          }
22      }
23      // TODO: defaultProps
24      PropsReactComp.defaultProps = {
25          default: "defaultProps"
26      };
27      // TODO: define const
28      const eleProps = <PropsReactComp />;
29      // TODO: React JSX
30      const reactSpan = (
31          <span>
32              <h3>React Props - defaultProps</h3>
33              {eleProps}
```

```
34            </span>
35        );
36        // TODO: React render
37        ReactDOM.render(reactSpan, divReact);
38    </script>
39 </body>
40 </html>
```

关于【代码 3-9】的说明：

- 第 18～22 行代码定义了一个 React 类组件（PropsReactComp），其中第 20 行代码中使用到了 props 参数（{this.props.default}）。读者一定注意到了，在声明类（PropsReactComp）的过程中明明没有定义 props 参数，那么这里如何使用 React Props 的呢？
- 关键是第 24～26 行代码的定义，通过 defaultProps 默认值为 React 类组件（PropsReactComp）定义了一个 "default" 属性（属性值为"defaultProps"）。这里定义好 defaultProps 默认值后，就可以像第 20 行代码那样（{this.props.default}）使用 React Props 了。

测试网页的效果如图 3.7 所示。如图中的箭头所示，通过 defaultProps 为 Props 设定默认值，可以为 React 类组件实现 React Props 应用。

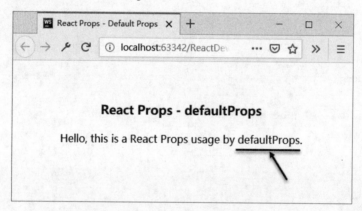

图 3.7　React Props 默认值

3.9　React 组件切分与提取

前面介绍了 React 组件和 React Props 的内容，相信读者已经开始慢慢习惯使用 React 组件设计了。通过 React 组件可以快速搭建出逻辑复杂的应用，但往往在初始设计过程中会形成功能模块不清晰、嵌套过多、很难复用的代码，这就需要设计人员后期对 React 组件进行切分，并提取出逻辑清晰、可高度复用的小组件，这样也有利于代码的修改便于维护。

下面，通过设计一个用户基本信息界面的代码实例，介绍 React 组件切分与提取的基本操作流程。

设计人员在刚开始构思一个用户基本信息界面时，往往会先把需求文档中的全部信息罗列进去，然后在页面中尽量完整体现业务逻辑所要求的功能。因此，初始代码设计大致会是下面的形式，具体如下：

【代码 3-10】（详见源代码目录 ch03-react-comp-extract-a.html 文件）

```
01  <!DOCTYPE html>
02  <html>
03  <head>
04      <meta charset="UTF-8"/>
05      <title>React Component - Extract</title>
06      <script src="https://unpkg.com/react@16/umd/react.development.js"></script>
07      <script src="https://unpkg.com/react-dom@16/umd/react-dom.development.js"></script>
08      <!-- Don't use this in production: -->
09      <script src="https://unpkg.com/babel-standalone@6.15.0/babel.min.js"></script>
10  </head>
11  <body>
12      <!-- 添加文档主体内容 -->
13      <div id='id-div-react'></div>
14      <script type="text/babel">
15          // TODO: get div
16          var divReact = document.getElementById('id-div-react');
17          // TODO: format date
18          function formatDate(date) {
19              return date.toLocaleDateString();
20          }
21          // TODO: composing component
22          function UserDetail(props) {
23              return (
24                  <div className="cssUserDetail">
25                      <span className="cssAvatar">
26                          <img className="cssAvaterImg"
27                              src={props.url}
28                              alt={props.alt}
29                          />
30                          <p className="p-center">Nickname: {props.nickname}</p>
31                      </span>
```

```
32              <span className="cssUserinfo">
33                  <p className="p-left">Id: {props.id}</p>
34                  <p className="p-left">Name: {props.name}</p>
35                  <p className="p-left">Gender: {props.gender}</p>
36                  <p className="p-left">Age: {props.age}</p>
37                  <p className="p-left">Email: {props.email}</p>
38              </span>
39              <span className="cssDate">
40                  <p className="p-right">Date: {formatDate(props.date)}</p>
41              </span>
42          </div>
43      );
44  }
45  // TODO: define const
46  const userDetail = <UserDetail
47      id="007"
48      name="King de. Wang"
49      nickname="Leo King"
50      gender="male"
51      age="18"
52      email="king@email.com"
53      url="images/avatar.png"
54      alt="loading..."
55      date={new Date()}
56  />;
57  // TODO: React JSX
58  const reactSpan = (
59      <span>
60          <h3>React Component - User Details</h3>
61          {userDetail}
62      </span>
63  );
64  // TODO: React render
65  ReactDOM.render(reactSpan, divReact);
66  </script>
67  </body>
68  </html>
```

关于【代码 3-10】的说明：

- 第 22～44 行代码定义了一个 React 函数组件（UserDetail），用于构成一个用户信息界面，具体说明如下：
  - 第 25～31 行代码通过<span>标签元素定义了第一个小面板，用于显示用户头像和用户昵称。

> 第 32~38 行代码通过<span>标签元素定义了第二个小面板，用于显示用户的详细信息（具体包括 id、name、gender、age 和 email 等）。

> 第 39~41 行代码通过<span>标签元素定义了第三个小面板，用于显示系统当前时间。

- 第 46~56 行代码通过 const 关键字定义了一个常量（UserDetail），赋值为 React 函数组件（<UserDetail>）。同时，定义了一组属性并初始化了属性值，通过 React Props 参数传递给 React 函数组件（UserDetail）。

我们注意到【代码 3-10】中整个用户信息界面的数据构成很杂乱，包括了图片地址、用户信息、日期时间等内容，且存在复杂的嵌套关系。这样的定义会造成设计人员阅读代码困难，后期代码修改和维护的难度也有所增加。

于是，React 组件的切分与提取就显得尤为重要了。其实，所谓的组件切分与提取就是将业务逻辑分类、归纳和精简。对于【代码3-10】而言，我们可以将整个用户信息界面切分为 3 个部分，分别是用户头像区域、用户信息区域和系统时间区域，修改后的代码如下：

【代码 3-11】（详见源代码目录 ch03-react-comp-extract-b.html 文件）

```
01  <!DOCTYPE html>
02  <html>
03  <head>
04      <meta charset="UTF-8"/>
05      <title>React Component - Extract</title>
06      <script src="https://unpkg.com/react@16/umd/react.development.js"></script>
07      <script src="https://unpkg.com/react-dom@16/umd/react-dom.development.js"></script>
08      <!-- Don't use this in production: -->
09      <script src="https://unpkg.com/babel-standalone@6.15.0/babel.min.js"></script>
10  </head>
11  <body>
12  <!-- 添加文档主体内容 -->
13  <div id='id-div-react'></div>
14  <script type="text/babel">
15      // TODO: get div
16      var divReact = document.getElementById('id-div-react');
17      // TODO: format date
18      function formatDate(date) {
19          return date.toLocaleDateString();
20      }
21      // TODO: composing component
22      function UserDetail(props) {
23          return (
```

```
24              <div className="cssUserDetail">
25                  <span className="cssAvatar">
26                      <img className="cssAvaterImg"
27                          src={props.avatar.url}
28                          alt={props.avatar.alt}
29                      />
30                      <p className="p-center">Nickname: {props.avatar.nickname}</p>
31                  </span>
32                  <span className="cssUserinfo">
33                      <p className="p-left">Id: {props.userinfo.id}</p>
34                      <p className="p-left">Name: {props.userinfo.name}</p>
35                      <p className="p-left">Gender: {props.userinfo.gender}</p>
36                      <p className="p-left">Age: {props.userinfo.age}</p>
37                      <p className="p-left">Email: {props.userinfo.email}</p>
38                  </span>
39                  <span className="cssDate">
40                      <p className="p-right">Date: {props.date.date}</p>
41                  </span>
42              </div>
43          );
44      }
45      // TODO: define const
46      const avatar = {
47          url: "images/avatar.png",
48          alt: "loading...",
49          nickname: "Leo King"
50      };
51      const userinfo = {
52          id: "007",
53          name: "King de. Wang",
54          gender: "male",
55          age: "18",
56          email: "king@email.com"
57      };
58      const date = {
59          date: formatDate(new Date())
60      };
61      const userDetail = <UserDetail
62          avatar={avatar}
63          userinfo={userinfo}
64          date={date}
```

```
65        />;
66        // TODO: React JSX
67        const reactSpan = (
68            <span>
69                <h3>React Component - User Details</h3>
70                {userDetail}
71            </span>
72        );
73        // TODO: React render
74        ReactDOM.render(reactSpan, divReact);
75    </script>
76  </body>
77  </html>
```

关于【代码 3-11】的说明：

- 关键部分是第 46～50 行代码、第 51～57 行代码和第 58～60 行代码的定义，分别定义了 "avatar" "userinfo" 和 "date" 3 个常量。这 3 个常量分别定义了用户头像、用户信息和系统时间的内容，然后使用在第 61～65 行代码定义的 React 函数组件（<userDetail>）中，并通过 React Props 参数进行传递。
- 第 22～44 行代码定义的 React 函数组件（UserDetail）与【代码 3-10】相比变化不大，主要区别是通过 Props 参数引用属性的方式不同。

我们注意到【代码 3-11】将全部用户信息划分成 3 个部分，分别是用户头像区域、用户信息区域和系统时间区域，但似乎代码逻辑还是很混乱。下面，我们尝试通过组件的方式将用户信息按照用户头像区域、用户信息区域和系统时间区域提取出来，分别构成 3 个小组件，然后再组合成一个大的组件，具体代码如下：

【代码 3-12】（详见源代码目录 ch03-react-comp-extract-c.html 文件）

```
01  <!DOCTYPE html>
02  <html>
03  <head>
04      <meta charset="UTF-8"/>
05      <title>React Component - Extract</title>
06      <script src="https://unpkg.com/react@16/umd/react.development.js"></script>
07      <script src="https://unpkg.com/react-dom@16/umd/react-dom.development.js"></script>
08      <!-- Don't use this in production: -->
09      <script src="https://unpkg.com/babel-standalone@6.15.0/babel.min.js"></script>
10  </head>
11  <body>
```

```
12      <!-- 添加文档主体内容 -->
13      <div id='id-div-react'></div>
14      <script type="text/babel">
15          // TODO: get div
16          var divReact = document.getElementById('id-div-react');
17          // TODO: format date
18          function formatDate(date) {
19              return date.toLocaleDateString();
20          }
21          // TODO: define const
22          const avatar = {
23              url: "images/avatar.png",
24              alt: "loading...",
25              nickname: "Leo King"
26          };
27          const userinfo = {
28              id: "007",
29              name: "King de. Wang",
30              gender: "male",
31              age: "18",
32              email: "king@email.com"
33          };
34          const date = {
35              date: formatDate(new Date())
36          };
37          // TODO: component Avatar
38          function Avatar(props) {
39              return (
40                  <span className="cssAvatar">
41                      <img className="cssAvaterImg"
42                          src={props.avatar.url}
43                          alt={props.avatar.alt}
44                      />
45                      <p className="p-center">Nickname: {props.avatar.nickname}</p>
46                  </span>
47              );
48          }
49          // TODO: component UserInfo
50          function UserInfo(props) {
51              return (
52                  <span className="cssUserinfo">
53                      <p className="p-left">Id: {props.userinfo.id}</p>
```

```
54              <p className="p-left">Name: {props.userinfo.name}</p>
55              <p className="p-left">Gender: {props.userinfo.gender}</p>
56              <p className="p-left">Age: {props.userinfo.age}</p>
57              <p className="p-left">Email: {props.userinfo.email}</p>
58          </span>
59      );
60  }
61  // TODO: composing component
62  function UserDetail(props) {
63      return (
64          <div className="cssUserDetail">
65              <Avatar avatar={props.avatar} />
66              <UserInfo userinfo={props.userinfo} />
67              <span className="cssDate">
68                  <p className="p-right">Date: {props.curdate.date}</p>
69              </span>
70          </div>
71      );
72  }
73  // TODO: define const userDetail
74  const userDetail = <UserDetail
75      avatar={avatar}
76      userinfo={userinfo}
77      curdate={date}
78  />;
79  // TODO: React JSX
80  const reactSpan = (
81      <span>
82          <h3>React Component - User Details</h3>
83          {userDetail}
84      </span>
85  );
86  // TODO: React render
87  ReactDOM.render(reactSpan, divReact);
88  </script>
89  </body>
90  </html>
```

关于【代码 3-12】的说明：

- 第 38~48 行代码定义的 React 函数组件（Avatar）和第 50~60 行代码定义的 React 函数组件（UserInfo）最为重要，注意到这两个函数组件的内容就是从【代码 3-11】定义的 React 函数组件（UserDetail）中所提取出来的。

- 在第62~72行代码定义的React函数组件（UserDetail）中，第65行和第66行代码分别引用了React函数组件（Avatar和UserInfo），具体来说就是通过小组件组合成了大组件。
- 另外，这段代码关于系统时间区域并没有按照组件切分的方式来实现，感兴趣的读者可以参照React函数组件（Avatar和UserInfo）的方式进行改写。

下面分别测试【代码3-10】、【代码3-11】和【代码3-12】所定义的HTML网页，其实3个网页的效果是完全一致的（读者可自行验证），具体如图3.8所示。

图3.8 React组件切分与提取

组件切分与提取看起来似乎是一项单调乏味的工作，但是在大型项目应用中，构建出逻辑清晰且可高度复用的组件是非常有价值的工作。当项目UI中有一部分模块重复使用了若干次，或者这部分模块自身就十分复杂，组件切分与提取就是一项十分有意义的工作。

第 4 章
◀ React State与生命周期 ▶

React 将组件看成是一个状态机（State Machines），通过其内部定义的状态（State）与生命周期（Lifecycle）实现并与用户的交互，维持组件不同的状态。本章将介绍 React State 与生命周期的内容。

4.1 React State 介绍

在 React 框架中定义了一个状态（State）概念，并通过状态（State）来实现 React 组件的状态机特性。所谓 React 组件的"状态机"特性，就是指组件通过与用户的交互，实现不同的状态，然后通过渲染 UI 保证用户界面和数据一致性。

React 框架之所以定义这个状态（State）概念，其目的就是仅仅通过更新 React 组件的状态（State），就可以实现重新渲染用户界面的操作（这样就不需要操作 DOM 了）。这点也正是 React 设计理念相较于其他前端框架的先进之处。

React State 的使用方法相比于 Props 较为复杂一些，但也是有基本模式的，下面就详细介绍一下。

首先，先要创建一个 ES6 形式的 React 组件（Component）类，这方面内容在前文中有过详细的介绍，具体代码如下：

【代码 4-1】

```
01  class ClassName extends React.Component {
02    // TODO: constructor
03    constructor() {}
04    // TODO: render
05    render() {
06      return ();
07    }
08  }
```

关于【代码 4-1】的说明：

- 在 React 组件类内部，要定义 constructor() 构造方法（见第 03 行代码），以及 render() 方法（见第 05～07 行代码）。

然后，在 render() 方法中通过 "this.state" 来使用 React 状态（State），具体的代码形式如下：

【代码 4-2】

```
01  class Clock extends React.Component {
02    // TODO: constructor
03    constructor() {}
04    // TODO: render
05    render() {
06      return (
07        <tagName>
08          {this.state...} // TODO: state usage
09        </tagName>
10      );
11    }
12  }
```

关于【代码 4-2】的说明：

- 第 08 行代码中，通过 "this.state" 来使用 React 状态（State）。

这里，React 状态（State）的初始化工作如何实现呢？这个工作需要放在 React 组件类的 constructor() 构造方法中，具体的代码形式如下：

【代码 4-3】

```
01  class Clock extends React.Component {
02    // TODO: constructor
03    constructor() {
04      this.state = {params: value};    // TODO: define & init state
05    }
06    // TODO: render
07    render() {
08      return (
09        <tagName>
10          {this.state...} // TODO: state usage
11        </tagName>
12      );
13    }
14  }
```

关于【代码 4-3】的说明：

- 第 04 行代码中，就是"this.state"的初始化方式。其中，"params"表示属性名称，"value"为初始化的属性值，同时可以定义若干组。

最后，还需要考虑将 Props 参数添加到构造方法中，具体的代码形式如下：

【代码 4-4】

```
01  class Clock extends React.Component {
02    // TODO: constructor
03    constructor(props) {
04      super(props);
05      this.state = {params: value};   // TODO: define & init state
06    }
07    // TODO: render
08    render() {
09      return (
10        <tagName>
11          {this.state...} // TODO: state usage
12        </tagName>
13      );
14    }
15  }
```

关于【代码 4-4】的说明：

- 第 03 行代码，在 constructor() 构造方法中定义 props 参数。
- 第 04 行代码，通过 super 关键字调用 props 参数，实现对父类构造方法的引用。这里的 super 关键字属于 JavaScript 语法范畴了。

通过以上几个基本步骤，就实现了 React State 的定义与使用，当然在实际应用中是很灵活的，需要多加练习才可以慢慢掌握。在下面的几个小节中，我们通过实现一个可封装重用的 React State 时钟应用来详细介绍 React 状态（State）的使用方法。

4.2 关于定时器时钟的思考

在本书第 1 章关于 React 渲染机制的介绍中，我们实现了一个时钟应用，具体是通过定时器机制来实现的。下面再回顾一下那段代码，具体如下：

【代码 4-5】（详见源代码目录 ch01-react-dom-render.html 文件）

```
01  <script type="text/babel">
02    /**
```

```
03        * update time
04        */
05       function updateTime() {
06           const renderDiv = (<div>
07               <h3>React 渲染机制</h3>
08               <p>现在时间是 {new Date().toLocaleTimeString()}.</p>
09           </div>);
10           // TODO: get div
11           var divReact = document.getElementById('id-div-react');
12           // TODO: render div
13           ReactDOM.render(renderDiv, divReact);
14       }
15       // TODO: define timer
16       setInterval(updateTime, 1000);
17   </script>
```

关于【代码 4-5】的说明：

- 第 05~14 行代码定义的 updateTime()方法，通过元素渲染的方式实现了 React 时钟效果（当然这个页面时钟是无法动态更新的）。
- 核心代码是第 16 行代码定义的计时器 setInterval()方法，在通过定时器定时调用 updateTime()方法，实现了页面时钟动态更新的效果。

虽然【代码 4-5】可以实现动态时钟的效果，但并不是我们预想中的完美方式。我们希望可以实现一个结构清晰、可封装和可重复使用的时钟组件。

4.3 开始封装时钟 UI

首先，我们就从封装 React State 时钟的 UI 开始，当然目前仅仅就是一个静态的时钟 UI，具体代码如下：

【代码 4-6】（详见源代码目录 ch04-react-state-clock-ui.html 文件）

```
01   <!DOCTYPE html>
02   <html>
03   <head>
04       <meta charset="UTF-8"/>
05       <title>React State - Clock</title>
06       <script src="https://unpkg.com/react@16/umd/react.development.js"></script>
07       <script src="https://unpkg.com/react-dom@16/umd/react-dom.development.js"></script>
08       <!-- Don't use this in production: -->
```

第 4 章 React State 与生命周期

```
09      <script
src="https://unpkg.com/babel-standalone@6.15.0/babel.min.js"></script>
10      </head>
11      <body>
12      <!-- 添加文档主体内容 -->
13      <div id='id-div-react'></div>
14      <script type="text/babel">
15          // TODO: get div
16          var divReact = document.getElementById('id-div-react');
17          // TODO: function component
18          function ClockReactComp() {
19              return <p>Hello, now is {new Date().toLocaleTimeString()}.</p>;
20          }
21          // TODO: React JSX
22          const reactSpan = (
23              <span>
24                  <h3>React State - Clock UI Wrapper</h3>
25                  <ClockReactComp />
26              </span>
27          );
28          // TODO: React render
29          ReactDOM.render(reactSpan, divReact);
30      </script>
31      </body>
32      </html>
```

关于【代码 4-6】的说明：

- 核心代码是第 18～20 行代码定义的函数组件（ClockReactComp()），实现了一个可封装的时钟 UI。目前，这个时钟 UI 是个静态组件，是无法自动更新的。

测试网页的效果如图 4.1 所示。如图中的箭头所示，页面中显示出了 React 时钟 UI 的样式，当然目前这个时钟是静态无法自动更新的。

图 4.1　React 时钟 UI

4.4 实现时钟 UI 的自动更新

在本节中，我们接着从前一节封装好 React State 时钟 UI 开始，实现时钟 UI 的自动更新功能，具体代码如下：

【代码 4-7】（详见源代码目录 ch04-react-state-clock-ui-auto.html 文件）

```html
01  <!DOCTYPE html>
02  <html>
03  <head>
04      <meta charset="UTF-8"/>
05      <title>React State - Clock</title>
06      <script src="https://unpkg.com/react@16/umd/react.development.js"></script>
07      <script src="https://unpkg.com/react-dom@16/umd/react-dom.development.js"></script>
08      <!-- Don't use this in production: -->
09      <script src="https://unpkg.com/babel-standalone@6.15.0/babel.min.js"></script>
10  </head>
11  <body>
12  <!-- 添加文档主体内容 -->
13  <div id='id-div-react'></div>
14  <script type="text/babel">
15      // TODO: get div
16      var divReact = document.getElementById('id-div-react');
17      // TODO: function component
18      function ClockReactComp(props) {
19          return <p>Hello, now is {props.date.toLocaleTimeString()}.</p>;
20      }
21      // TODO: auto Clock
22      function autoClock() {
23          ReactDOM.render(
24              <span>
25                  <h3>React State - Clock UI Wrapper Update</h3>
26                  <ClockReactComp date={new Date()} />
27              </span>
28              , divReact);
29      }
30      setInterval(autoClock, 1000); // TODO: set timer
31  </script>
32  </body>
```

```
33    </html>
```

关于【代码 4-7】的说明：

- 首先，我们需要完整保留第 18~20 行代码定义的时钟函数组件（ClockReactComp()），目的是保留对时钟 UI 的封装。
- 然后，我们会想到通过定义计时器实现时钟的自动更新，这样就添加了第 30 行代码定义的 setInterval()方法，目的是通过这个定时器方法每间隔 1 秒来更新一次时钟的显示。由于 setInterval()方法的语法特点，我们需要定义一个回调函数 autoClock()。
- 最后，第 22~29 行代码是 autoClock()函数的实现过程，通过调用时钟函数组件（ClockReactComp()）来显示页面时钟，并借助 Props 参数来传递时间。

测试网页的效果如图 4.2 所示。如图中箭头所示，页面中显示出的 React 时钟 UI 已经可以自动更新了。其实，【代码 4-7】本质上与【代码 4-5】是类似的，只不过突出了可封装的时钟组件的概念。

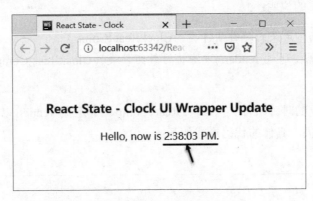

图 4.2　React 时钟 UI 的自动更新

4.5　自我更新的时钟 UI 类

在前一节中，我们实现了可封装的 React 时钟 UI 的自动更新，主要是通过定时器功能实现的。不过，对于完美主义者还是不够理想，与 React 框架的设计理念也有很大的差距。

我们理想的形式是仅仅定义一个时钟组件，该时钟组件可以自我更新，将其渲染到页面中就是一个真正的时钟 UI。因此，具体的代码形式大致如下：

【代码 4-8】

```
01  ReactDOM.render(
02    <Clock />,
03    document.getElementById('div')
04  );
```

关于【代码 4-8】的说明：

- 第 04 行代码定义的 <Clock /> 就是我们想要的、可封装、可重用的时钟组件。这个时钟组件在渲染到页面中之后，是完全可以自我更新的。

为了实现这个可自我更新的时钟 UI，我们需要将函数组件转换为 ES6 类组件，这个类组件具有自我更新的功能。

（1）需要创建一个 ES6 组件类，并且继承于 React.Component，具体的代码形式如下：

【代码 4-9】

```
01  // TODO: define ES6 Class Component
02  class ClockReactComp extends React.Component {...}
03  // TODO: React render
04  ReactDOM.render(<ClockReactComp />, divReact);
```

关于【代码 4-9】的说明：

- 第 02 行代码，定义的时钟组件类（ClockReactComp）。
- 第 04 行代码，通过 render() 方法将时钟组件类（ClockReactComp）渲染到页面中进行显示。

（2）在刚刚创建的时钟组件类（ClockReactComp）中定义 render() 方法，并将之前时钟函数组件的内容加进去，具体的代码形式如下：

【代码 4-10】

```
01  // TODO: define ES6 Class Component
02  class ClockReactComp extends React.Component {
03      render() {
04          return (
05              <span>
06                  <h3>React State - Clock Class</h3>
07                  <p>Hello, now is {this.props.date.toLocaleTimeString()}.</p>
08              </span>
09          );
10      }
11  }
12  // TODO: React render
13  ReactDOM.render(<ClockReactComp />, divReact);
```

关于【代码 4-10】的说明：

- 第 03～10 行代码，通过 render() 方法定义的、所渲染的时钟函数组件的内容。
- 另外，由于在 ES6 组件类中无法显式地定义 Props 参数，需要在 props 参数前使用 this 关键字引用。

（3）到这步为止，一个可自我更新的时钟组件类（ClockReactComp）的雏形就基本完成了，然后借助【代码 4-7】的机制通过定时器来实现时钟的自我更新，完整的代码形式如下：

【代码 4-11】（详见源代码目录 ch04-react-state-clock-class.html 文件）

```
01    <!DOCTYPE html>
02    <html>
03    <head>
04        <meta charset="UTF-8"/>
05        <title>React State - Clock</title>
06        <script src="https://unpkg.com/react@16/umd/react.development.js"></script>
07        <script src="https://unpkg.com/react-dom@16/umd/react-dom.development.js"></script>
08        <!-- Don't use this in production: -->
09        <script src="https://unpkg.com/babel-standalone@6.15.0/babel.min.js"></script>
10    </head>
11    <body>
12    <!-- 添加文档主体内容 -->
13    <div id='id-div-react'></div>
14    <script type="text/babel">
15        // TODO: get div
16        var divReact = document.getElementById('id-div-react');
17        // TODO: define ES6 Class Component
18        class ClockReactComp extends React.Component {
19            render() {
20                return (
21                    <span>
22                        <h3>React State - Clock Class</h3>
23                        <p>Hello, now is {this.props.date.toLocaleTimeString()}.</p>
24                    </span>
25                );
26            }
27        }
28        // TODO: auto Clock
29        function autoClock() {
30            ReactDOM.render(
31                <ClockReactComp date={new Date()} />,
32                divReact
33            );
34        }
35        // TODO: set timer
```

```
36          setInterval(autoClock, 1000);
37     </script>
38   </body>
39 </html>
```

关于【代码 4-11】的说明：

- 第 29~34 行代码是 autoClock() 函数的实现过程，通过调用时钟组件类（ClockReactComp()）来显示页面时钟，并借助 Props 参数来传递时间。
- 在第 36 行代码中，通过定义计时器来实现时钟组件类（ClockReactComp()）的自我更新。

测试网页的效果如图 4.3 所示。如图中的箭头所示，通过可自我更新的时钟组件类也可以实现同样的效果。同时，关键的地方是【代码 4-11】通过 ES6 Class 形式定义的时钟组件，更接近我们理想中的 React 时钟组件了。

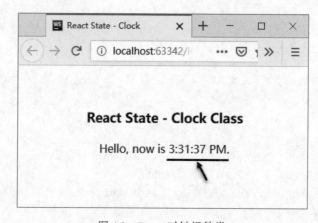

图 4.3 React 时钟组件类

4.6 在时钟组件类中引入 State 状态

在这一节中，我们开始正式引入 React 状态（State）的概念。前面做了那么多的代码铺垫，就是为了在这里引入 State 状态的概念时能够水到渠成，因为 React State 是其他大部分 JavaScript 前端框架中所没有的概念。

为了便于读者学习，我们就以【代码 4-11】为代码蓝本，通过代码改写的方式进行介绍。另外，改写的方式可以参考本章第 1 小节中介绍的内容。

首先，我们需要将 "props" 参数替换为 "state" 属性，具体代码形式如下：

【代码 4-12】（详见源代码目录 ch04-react-state-clock-state.html 文件）

```
01 class ClockReactComp extends React.Component {
```

```
02    render() {
03        return (
04            <span>
05                <h3>React State - Clock Class State</h3>
06                <p>Hello, now is {this.state.date.toLocaleTimeString()}.</p>
07            </span>
08        );
09    }
10 }
```

关于【代码 4-12】的说明：

- 第 06 行代码中，使用"state"状态替换了原来的"props"参数。

接着，需要为时钟组件类（ClockReactComp）添加 constructor()构造方法，并且对"state"状态进行初始化，具体代码形式如下：

【代码 4-13】（详见源代码目录 ch04-react-state-clock-state.html 文件）

```
01 class ClockReactComp extends React.Component {
02     constructor() {
03         this.state = {date: new Date()};
04     }
05     render() {
06         return (
07             <span>
08                 <h3>React State - Clock Class State</h3>
09                 <p>Hello, now is {this.state.date.toLocaleTimeString()}.</p>
10             </span>
11         );
12     }
13 }
```

关于【代码 4-13】的说明：

- 第 02～04 行代码定义的就是时钟组件类（ClockReactComp）的 constructor()构造方法。其中，第 03 行代码对"state"状态进行了初始化操作，定义了一个 date 时间属性。

然后，将"props"参数传递到父类的构造方法之中，具体的代码形式如下：

【代码 4-14】（详见源代码目录 ch04-react-state-clock-state.html 文件）

```
01 class ClockReactComp extends React.Component {
02     constructor(props) {
03         super(props);
04         this.state = {date: new Date()};
05     }
```

```
06        render() {
07            return (
08                <span>
09                    <h3>React State - Clock Class State</h3>
10                    <p>Hello, now is {this.state.date.toLocaleTimeString()}.</p>
11                </span>
12            );
13        }
14    }
```

关于【代码 4-14】的说明:

- 在第 03 行代码中,通过 super 关键字调用 "props" 参数,实现将 "props" 参数传递到父类构造方法的操作。

至此,在时钟组件类(ClockReactComp)中引入 React State 状态的操作就基本完成了,完整的代码形式如下:

【代码 4-15】(详见源代码目录 ch04-react-state-clock-state.html 文件)

```
01  <!DOCTYPE html>
02  <html>
03  <head>
04      <meta charset="UTF-8"/>
05      <title>React State - Clock</title>
06      <script src="https://unpkg.com/react@16/umd/react.development.js"></script>
07      <script src="https://unpkg.com/react-dom@16/umd/react-dom.development.js"></script>
08      <!-- Don't use this in production: -->
09      <script src="https://unpkg.com/babel-standalone@6.15.0/babel.min.js"></script>
10  </head>
11  <body>
12  <!-- 添加文档主体内容 -->
13  <div id='id-div-react'></div>
14  <script type="text/babel">
15      // TODO: get div
16      var divReact = document.getElementById('id-div-react');
17      // TODO: define ES6 Class Component
18      class ClockReactComp extends React.Component {
19          constructor(props) {
20              super(props);
21              this.state = {date: new Date()};
22          }
```

```
23          render() {
24              return (
25                  <span>
26                      <h3>React State - Clock Class State</h3>
27                      <p>Hello, now is {this.state.date.toLocaleTimeString()}.</p>
28                  </span>
29              );
30          }
31      }
32      // TODO: React render
33      ReactDOM.render(<ClockReactComp />, divReact);
34  </script>
35  </body>
36  </html>
```

关于【代码4-15】的说明：

- 在第 33 行代码中，通过调用 ReactDOM 的 render()方法，实现将时钟组件类（ClockReactComp()）渲染到页面中进行显示的操作。另外请注意，我们在 <ClockReactComp />元素中移除了 date 属性。

测试网页的效果如图 4.4 所示。通过在时钟组件类（ClockReactComp()）中引入 State 状态，也可以实现同样的页面显示效果。正如图中标识所写内容那样，这个时钟组件类是无法自我更新的。难道还是要继续通过计时器功能，才可以实现自我更新的时钟组件类么？我们在下一小节中给出答案。

图 4.4　React State 时钟组件类

4.7 React 生命周期介绍

在前一节的结尾,我们提出了一个问题。下面就告诉读者,React 框架为组件设计了一个"生命周期"的概念,用于配合 React 状态(State)实现组件的渲染操作。

在 React 组件中,生命周期可基本分成三个状态,具体如下:

- Mounting:已开始挂载真实的组件 DOM。
- Updating:正在重新渲染组件 DOM。
- Unmounting:已卸载真实的组件 DOM。

同时,React 框架定义了一组关于生命周期的方法,具体如下:

- componentWillMount()方法:在渲染前调用,可以在客户端,也可以在服务端。
- componentDidMount()方法:在第一次渲染后调用,只作用于客户端。
- componentWillUpdate()方法:在组件接收到新的 Props 参数或者 State 状态、但还没有渲染时被调用。另外,该方法在初始化时不会被调用。
- componentDidUpdate()方法:在组件完成更新后会立即调用。另外,该方法在初始化时不会被调用。
- componentWillUnmount()方法:在组件从 DOM 中被移除之前会立刻被调用。

以上这组与生命周期相关的方法,可以放到 ES6 组件类中进行使用,从而实现对 React 组件状态的控制。下面在【代码 4-4】的基础上进行简单的修改,演示一下这组方法的用法:

【代码 4-16】

```
01  class Clock extends React.Component {
02      // TODO: constructor
03      constructor(props) {
04          super(props);
05          this.state = {params: value};// TODO: define & init state
06      }
07      // TODO: realize lifecycle methods
08      componentDidMount() {}
09      componentWillUnmount() {}
10      // TODO: render
11      render() {
12          return (
13              <tagName>
14                  {this.state...} // TODO: state usage
15              </tagName>
16          );
17      }
```

```
18    }
```

关于【代码4-16】的说明：

- 第08行和第09行代码所定义的方法，就是对React生命周期的使用过程。当然，设计人员可以选择自己项目所需的生命周期方法，无需将全部生命周期方法都放进组件类中。

4.8 在时钟组件类中使用生命周期

在这一节中，我们通过引入React生命周期的概念，配合State状态来实现最终版本的React时钟组件。

我们继续以【代码4-15】为基础，通过引入相关生命周期方法来实现能够自我更新的时钟组件类，具体代码如下：

【代码4-17】（详见源代码目录 ch04-react-state-clock-lifecycle.html 文件）

```
01  <!DOCTYPE html>
02  <html>
03  <head>
04      <meta charset="UTF-8"/>
05      <title>React State - Clock</title>
06      <script src="https://unpkg.com/react@16/umd/react.development.js"></script>
07      <script src="https://unpkg.com/react-dom@16/umd/react-dom.development.js"></script>
08      <!-- Don't use this in production: -->
09      <script src="https://unpkg.com/babel-standalone@6.15.0/babel.min.js"></script>
10  </head>
11  <body>
12  <!-- 添加文档主体内容 -->
13  <div id='id-div-react'></div>
14  <script type="text/babel">
15      // TODO: get div
16      var divReact = document.getElementById('id-div-react');
17      // TODO: define ES6 Class Component
18      class ClockReactComp extends React.Component {
19          // TODO: constructor
20          constructor(props) {
21              super(props);
22              this.state = {date: new Date()};
23          }
```

```
24          // TODO: Lifecycle methods
25          componentDidMount() {
26            this.timerId = setInterval(
27              () => this.tick(),
28              1000
29            );
30          }
31          componentWillUnmount() {
32            clearInterval(this.timerId);
33          }
34          // TODO: update state
35          tick() {
36            this.setState({
37              date: new Date()
38            });
39          }
40          // TODO: render
41          render() {
42            return (
43              <span>
44                <h3>React State - Clock Class Lifecycle</h3>
45                <p>Hello, now is {this.state.date.toLocaleTimeString()}.</p>
46              </span>
47            );
48          }
49        }
50        // TODO: React render
51        ReactDOM.render(
52          <ClockReactComp/>,
53          divReact
54        );
55    </script>
56  </body>
57 </html>
```

关于【代码 4-17】的说明：

- 在第 25～30 行代码中引入了 componentDidMount()方法，该方法会在组件已经被渲染到 DOM 中后被触发，可以在该方法中设置一个计时器，来实现时钟组件类的自我更新功能。其中，第 26～29 行代码通过使用 setInterval()定义了一个计时器（this.timerId），并通过该定时器定时调用第 27 行代码所定义的箭头函数方法（tick()）。

- 第35～39行代码是自定义方法（tick()）的实现过程，主要是通过 this.setState()方法来更新组件的 State 状态。其中，第37行代码通过更新 date 属性实现了时钟组件类的自我更新。
- 在第31～33行代码中引入了 componentWillUnmount()方法，该方法会在组件已经被从 DOM 中清除后被触发。其中，第32行代码通过使用 clearInterval()清除了计时器（this.timerId）。

测试网页的效果如图 4.5 所示。通过在时钟组件类（ClockReactComp()）中定义 State 状态和使用生命周期方法调用定时器，同样实现了的自我更新的时钟 UI 效果。同时，该时钟组件类（ClockReactComp()）通过结合 State 状态和使用生命周期方法，完全实现了一个结构清晰、可封装和可重复使用的时钟组件。

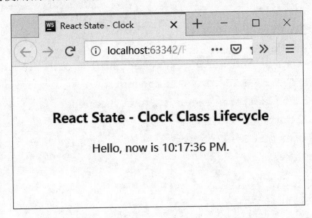

图 4.5　React State & Lifecycle 时钟组件类

4.9　正确的使用 State 状态

在 React 框架中，对于如何正确地使用状态（State）定义了专属的方式，具体就是通过 setState()方法来更新状态（State）。

这里需要强调的是，在 React 框架中是不允许直接操作状态（State）属性的，而且直接操作是无效的。如何理解呢？简单说就是直接操作状态（State）属性时，虽然 React 框架不会报错，但是不会产生任何效果。

下面，我们先通过一个具体的代码实例讲解一下，如果直接操作状态（State）属性会是什么效果，具体代码如下：

【代码 4-18】（详见源代码目录 ch04-react-state-direct-change.html 文件）

```
01  <!DOCTYPE html>
02  <html>
03  <head>
```

```
04      <meta charset="UTF-8"/>
05      <title>React State - State</title>
06      <script src="https://unpkg.com/react@16/umd/react.development.js"></script>
07      <script src="https://unpkg.com/react-dom@16/umd/react-dom.development.js"></script>
08      <!-- Don't use this in production: -->
09      <script src="https://unpkg.com/babel-standalone@6.15.0/babel.min.js"></script>
10    </head>
11    <body>
12    <!-- 添加文档主体内容 -->
13    <div id='id-div-react'></div>
14    <script type="text/babel">
15        // TODO: get div
16        var divReact = document.getElementById('id-div-react');
17        // TODO: define ES6 Class Component
18        class CountReactComp extends React.Component {
19            constructor(props) {
20                super(props);
21                this.state = {
22                    count: 0
23                };
24            }
25            componentDidMount() {
26                this.timerId = setInterval(
27                    () => this.count(),
28                    1000
29                );
30            }
31            count() {
32                this.state.count += this.props.increment;
33            }
34            componentWillUnmount() {
35                clearInterval(this.timerId);
36            }
37            render() {
38                return (
39                    <span>
40                        <h3>React State - Direct Change State</h3>
41                        <p>Hello, count is {this.state.count}.</p>
42                    </span>
43                );
```

```
44          }
45      }
46      // TODO: defaultProps
47      CountReactComp.defaultProps = {
48          increment: 1
49      };
50      // TODO: React render
51      ReactDOM.render(<CountReactComp/>, divReact);
52  </script>
53  </body>
54  </html>
```

关于【代码 4-18】的说明：

- 在第 19~24 行代码定义的构造方法中，定义了一个状态（State）属性（count），并初始化为数值 0，该属性用于测试直接操作的方式。
- 在第 25~30 行代码定义的 componentDidMount() 方法中，通过使用 setInterval() 方法定义了一个计时器（this.timerId），并通过箭头函数调用了 count() 方法。
- 第 31~33 行代码是自定义方法（count()）的实现过程，通过直接操作"count"属性进行累加（+1）。其中，props 参数（increment）定义在第 47~49 行代码中。
- 在第 41 行代码中，通过 render() 方法将状态（State）属性（count）的值渲染到页面中进行显示。
- 在第 34~36 行代码定义的 componentWillUnmount() 方法中，通过使用 clearInterval() 方法清除了上面定义的计时器（this.timerId）。

测试网页的效果如图 4.6 所示。如图中的箭头所示，在第 32 行代码中尝试直接通过操作"count"属性进行累加（+1），页面中是看不到任何的效果，这说明该方式并不会重新渲染组件。

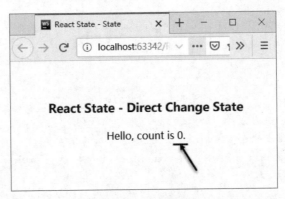

图 4.6 React 直接操作 State 属性

下面，我们通过另一个代码实例，讲解一下通过 setState() 方法操作状态（State）属性的正确方式，具体代码如下：

【代码 4-19】（详见源代码目录 ch04-react-state-setState.html 文件）

```html
01  <!DOCTYPE html>
02  <html>
03  <head>
04      <meta charset="UTF-8"/>
05      <title>React State - setState</title>
06      <script src="https://unpkg.com/react@16/umd/react.development.js"></script>
07      <script src="https://unpkg.com/react-dom@16/umd/react-dom.development.js"></script>
08      <!-- Don't use this in production: -->
09      <script src="https://unpkg.com/babel-standalone@6.15.0/babel.min.js"></script>
10  </head>
11  <body>
12  <!-- 添加文档主体内容 -->
13  <div id='id-div-react'></div>
14  <script type="text/babel">
15      // TODO: get div
16      var divReact = document.getElementById('id-div-react');
17      // TODO: define ES6 Class Component
18      class CountReactComp extends React.Component {
19          constructor(props) {
20              super(props);
21              this.state = {
22                  count: 0
23              };
24          }
25          componentDidMount() {
26              this.timerId = setInterval(
27                  () => this.count(),
28                  1000
29              );
30          }
31          count() {
32              this.setState((prevState, props) => ({
33                  coun: prevState.count + props.increment
34              }));
35          }
36          componentWillUnmount() {
37              clearInterval(this.timerId);
38          }
39          render() {
```

```
40              return (
41                  <span>
42                      <h3>React State - setState()</h3>
43                      <p>Hello, count is {this.state.count}.</p>
44                  </span>
45              );
46          }
47      }
48      // TODO: defaultProps
49      CountReactComp.defaultProps = {
50          increment: 1
51      };
52      // TODO: React render
53      ReactDOM.render(<CountReactComp/>, divReact);
54  </script>
55  </body>
56  </html>
```

关于【代码 4-19】的说明：

- 这段代码与【代码 4-18】基本类似，不同之处就是第 31~35 行代码所实现的自定义方法（count()）。
- 在自定义方法（count()）中，第 32~34 行代码通过 setState()方法操作"count"属性进行累加（+1）。
- 另外，在 setState()方法中定义了两个参数"prevState"和"props"，分别用于传递前一个 State 状态和 Props 参数。

测试网页的效果如图 4.7 所示。如图中的箭头所示，通过 setState()方法操作状态（State）属性的正确方式，可以看到数值动态更新的效果。

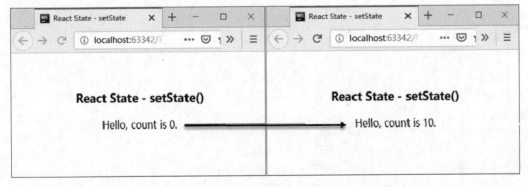

图 4.7　React – setState()方法

4.10 自顶向下的数据流

从前文中知道，React 框架中的组件是被定义为具有状态（State）的。但是，无论是父组件或子组件都不能知道某个组件是有状态还是无状态的，并且自己也并不知道（不应该去关心）某个组件是通过函数方式或是通过 ES6 Class 方式去定义的。

这点恰好解释了"React 状态（State）通常被称为局部的或封装的"的原因。在 React 框架中，状态（State）除了定义它的自身组件，其他的组件均是不可访问它的。因此，无论是对于父组件或是子组件，都无法知道某个组件是有状态的还是无状态的，也并不知道（不应该去关心）某个组件是函数组件或是 ES6 Class 组件。

在 React 框架中，可以将组件所定义的状态（State）作为 Props 参数向下传递到其子组件中，但是子组件却无法知道该参数是来自于父组件的状态（State）、参数（Props）或者是人工输入的。

这一点如何理解呢？下面，我们通过一个具体的代码实例讲解一下：

【代码 4-20】（详见源代码目录 ch04-react-state-data-flow.html 文件）

```
01  <!DOCTYPE html>
02  <html>
03  <head>
04      <meta charset="UTF-8"/>
05      <title>React State - Data Flow</title>
06      <script src="https://unpkg.com/react@16/umd/react.development.js"></script>
07      <script src="https://unpkg.com/react-dom@16/umd/react-dom.development.js"></script>
08      <!-- Don't use this in production: -->
09      <script src="https://unpkg.com/babel-standalone@6.15.0/babel.min.js"></script>
10  </head>
11  <!-- 添加文档主体内容 -->
12  <div id='id-div-react'></div>
13  <script type="text/babel">
14      // TODO: get div
15      var divReact = document.getElementById('id-div-react');
16      // TODO: function component
17      function FormattedDate(props) {
18          return <h3>Now is {props.date.toLocaleTimeString()}.</h3>;
19      }
20      // TODO: define ES6 Class Component
21      class ClockReactComp extends React.Component {
22          static defaultProps = {
```

```
23            propsDate: new Date()
24        };
25        constructor(props) {
26            super(props);
27            this.state = {date: new Date()};
28        }
29        componentDidMount() {
30            this.timerId = setInterval(
31                () => this.tick(),
32                1000
33            );
34        }
35        componentWillUnmount() {
36            clearInterval(this.timerId);
37        }
38        tick() {
39            this.setState({
40                date: new Date()
41            });
42        }
43        render() {
44            return (
45                <span>
46                    <h3>React State - Data Flow Clock</h3>
47                    <FormattedDate date={this.state.date} />
48                    <FormattedDate date={new Date()} />
49                    <FormattedDate date={this.props.propsDate} />
50                </span>
51            );
52        }
53    }
54    // TODO: React render
55    ReactDOM.render(<ClockReactComp />, divReact);
56 </script>
57 </body>
58 </html>
```

关于【代码 4-20】的说明：

- 第 17~19 行代码定义了一个函数组件（FormattedDate），用于在页面中显示当前时间。
- 第 21~53 行代码定义了一个类组件（ClockReactComp），相当于一个时钟组件。其中，在第 47~49 行代码所定义的函数组件（FormattedDate）渲染方法中，分别使用了三种方式。

- 第 47 行代码是通过类组件（ClockReactComp）的状态属性（date）获取的时间，通过 Props 参数传递给函数组件（FormattedDate）的方式渲染的。
- 第 48 行代码是通过直接获取时间，通过 Props 参数传递给函数组件（FormattedDate）的方式渲染的。
- 第 49 行代码是通过类组件（ClockReactComp）的 Props 参数（propsDate）获取的时间，通过 Props 参数传递给函数组件（FormattedDate）的方式渲染的。其中，参数"propsDate"的定义在第 22~24 行代码中。

测试网页的效果如图 4.8 所示。页面中同时显示出了 3 个时钟的效果，说明函数组件（FormattedDate）是不关心时间参数是如何从时钟组件（ClockReactComp）获取的，也不关心时钟组件（ClockReactComp）到底是函数组件还是类组件。

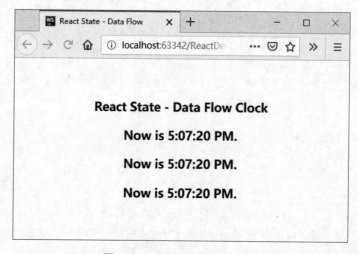

图 4.8　React Data Flow（一）

在 React 框架中，这被称为"自顶而下"的数据流，就好比树形结构所构成的、自顶而下的单向数据流。对于 React 状态（State）而言，其仅仅属于自身的组件，而且从该状态（State）所派生的任何数据或 UI 只能影响其所派生的组件。

React 官方文档中对此有一个形象的比喻——"数据瀑布"。具体解释就是，在一个以组件为节点所构成的树形瀑布中，每一个组件的状态（State）就像在此节点上额外增加的水源，仅仅属于该节点，并且只能向下单向流动。

下面，我们在【代码 4-20】的基础上稍做修改，创建一个同时渲染三个时钟组件的 App 组件，借此说明一下"数据瀑布"的特性，具体代码如下：

【代码 4-21】（详见源代码目录 ch04-react-state-data-flow-tri.html 文件）

```
01  <!DOCTYPE html>
02  <html>
03  <head>
04      <meta charset="UTF-8"/>
```

```
05      <title>React State - Data Flow</title>
06      <script src="https://unpkg.com/react@16/umd/react.development.js"></script>
07      <script src="https://unpkg.com/react-dom@16/umd/react-dom.development.js"></script>
08      <!-- Don't use this in production: -->
09      <script src="https://unpkg.com/babel-standalone@6.15.0/babel.min.js"></script>
10    </head>
11    <body>
12      <!-- 添加文档主体内容 -->
13      <div id='id-div-react'></div>
14      <script type="text/babel">
15        // TODO: get div
16        var divReact = document.getElementById('id-div-react');
17        // TODO: function component
18        function FormattedDate(props) {
19            return <h3>Now is {props.date.toLocaleTimeString()}.</h3>;
20        }
21        // TODO: define ES6 Class Component
22        class ClockReactComp extends React.Component {
23            static defaultProps = {
24                propsDate: new Date()
25            };
26            constructor(props) {
27                super(props);
28                this.state = {date: new Date()};
29            }
30            componentDidMount() {
31                this.timerId = setInterval(
32                    () => this.tick(),
33                    1000
34                );
35            }
36            componentWillUnmount() {
37                clearInterval(this.timerId);
38            }
39            tick() {
```

```
40          this.setState({
41              date: new Date()
42          });
43       }
44       render() {
45          return (
46              <span>
47                  <h3>React State - Data Flow Clock App</h3>
48                  <FormattedDate date={this.state.date} />
49              </span>
50          );
51       }
52   }
53   // TODO: define app
54   function ClockDataFlow() {
55       return (
56          <div>
57              <ClockReactComp />
58              <ClockReactComp />
59              <ClockReactComp />
60          </div>
61       );
62   }
63   // TODO: React render
64   ReactDOM.render(<ClockDataFlow />, divReact);
65   </script>
66   </body>
67   </html>
```

关于【代码 4-21】的说明：

- 第 54～62 行代码增加定义了一个组件（ClockDataFlow），其中第 57～59 行代码同时引用了 3 个相同的时钟组件（ClockReactComp）。

测试网页的效果如图 4.9 所示。3 个时钟组件（ClockReactComp）每个都单独设置其自己的计时器、并且独立进行更新。

第 4 章 React State 与生命周期

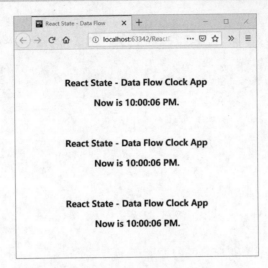

图 4.9　React Data Flow（二）

由于【代码 4-21】所定义的 3 个时钟组件（ClockReactComp）是同时启动计时器的，也许时钟独立动态更新的效果不明显。下面，我们尝试再将【代码 4-21】稍做修改，让 3 个时钟组件表现出差异性，具体代码如下：

【代码 4-22】（详见源代码目录 ch04-react-state-data-flow-tri-random.html 文件）

```
01  <!DOCTYPE html>
02  <html>
03  <head>
04      <meta charset="UTF-8"/>
05      <title>React State - Data Flow</title>
06      <script src="https://unpkg.com/react@16/umd/react.development.js"></script>
07      <script src="https://unpkg.com/react-dom@16/umd/react-dom.development.js"></script>
08      <!-- Don't use this in production: -->
09      <script src="https://unpkg.com/babel-standalone@6.15.0/babel.min.js"></script>
10  </head>
11  <body>
12  <!-- 添加文档主体内容 -->
13  <div id='id-div-react'></div>
14  <script type="text/babel">
15      // TODO: get div
16      var divReact = document.getElementById('id-div-react');
17      // TODO: function component
18      function FormattedDate(props) {
19          return <h3>Now is {props.date.toLocaleTimeString()}.</h3>;
20      }
```

```
21      // TODO: define ES6 Class Component
22      class ClockReactComp extends React.Component {
23          static defaultProps = {
24              propsDate: new Date()
25          };
26          constructor(props) {
27              super(props);
28              this.state = {date: new Date()};
29          }
30          componentDidMount() {
31              this.timerId = setInterval(
32                  () => this.tick(),
33                  Math.ceil(Math.random()*9)*1000
34              );
35          }
36          componentWillUnmount() {
37              clearInterval(this.timerId);
38          }
39          tick() {
40              this.setState({
41                  date: new Date()
42              });
43          }
44          render() {
45              return (
46                  <span>
47                      <h3>React State - Data Flow Clock Random App</h3>
48                      <FormattedDate date={this.state.date} />
49                  </span>
50              );
51          }
52      }
53      // TODO: define app
54      function ClockDataFlow() {
55          return (
56              <div>
57                  <ClockReactComp />
58                  <ClockReactComp />
59                  <ClockReactComp />
60              </div>
61          );
62      }
63      // TODO: React render
```

```
64    ReactDOM.render(<ClockDataFlow />, divReact);
65  </script>
66  </body>
67  </html>
```

关于【代码 4-22】的说明：

- 第 31～34 行代码定义的 setInterval()计时器方法中，第 33 行代码是通过随机函数生成的时间间隔。目的很简单，在 3 个时钟组件初始化的过程中，每个时钟组件更新的时间间隔就是随机的。因此，3 个时钟组件在页面渲染出来的效果（更新时间间隔）是不同的。

测试网页的效果如图 4.10 所示。3 个时钟组件（ClockReactComp）所显示的时间是不同的，说明每个时钟组件定义的计时间隔是随机的。这一点更好地证明了每个时钟组件都是单独设置其自己的计时器，并且独立地进行更新。

图 4.10　React Data Flow（三）

第 5 章

◀ React 事件处理 ▶

React 框架的事件处理机制与 JavaScript 的 DOM 元素事件类似，但二者在语法上略有不同。本章将详细为读者介绍 React 框架的事件处理。

5.1 React 事件介绍

在 React 框架中，React 元素的事件处理和 JavaScript 对 HTML DOM 元素处理的方式类似。但是，二者在语法上略有不同，根据官方文档的说明描述如下：

- React 事件绑定属性的命名采用驼峰式写法，而不是小写。
- 如果采用 JSX 语法，则需要传入一个函数作为事件处理函数，而不是一个字符串（DOM 元素的写法）。

那么，对于上面的两条规定如何理解呢？下面，我们通过对比 HTML DOM 事件语法和 React 事件语法的书写方法，具体解释一下这两条规定。

比如最常见的按钮（<button>）鼠标单击事件（Click），在 HTML DOM 事件语法中需要写成如下的方式：

【代码 5-1】

```
// TODO: HTML DOM Event
<button onclick="on_btn_click()">
Click Me!
</button>
```

关于【代码 5-1】的说明：

- onclick 是鼠标单击事件名称，注意事件名称全部是小写的。

on_btn_click() 是鼠标单击事件的函数方法，注意需要用双引号(" ")将函数方法包括进去；

相信以上内容对于大多数读者来讲应该很熟悉，如果写 JavaScript 事件处理的代码就很自然的能理解。而如果【代码 5-1】使用 React 框架来实现，该如何写呢？请继续看下面的代码：

【代码 5-2】

```
// TODO: React Event
<button onClick={on_btn_click}>
Click Me!
</button>
```

关于【代码 5-2】的说明：

- onClick 是鼠标单击事件名称，注意事件名称是采用小驼峰（注意小驼峰写法和大驼峰写法的区别）写法的。
- on_btn_click 是鼠标单击事件的函数方法，注意是采用 JSX 方式（{ }）将函数方法包括进去的，而且不带小括号。

上面就是 React 框架中对于事件语法的定义，与传统的 HTML DOM 事件语法是略有区别的。

另外，在 React 框架中对于阻止事件默认行为的方式也是不同的，传统方式下可以使用返回 "false" 来阻止默认行为，但在 React 框架中则必需显式的使用 preventDefault()方法来阻止默认行为。

比如在传统的 HTML 页面中，如果想阻止链接（<a>）默认打开一个新页面，可以这样书写：

【代码 5-3】

```
// TODO: HTML DOM Event
<a href="#" onclick="alert('Prevent hyperlink address open new page!'); return false">
   Click Me!
</a>
```

关于【代码 5-3】的说明：

- 通过返回 "false"（return false）就可以实现阻止链接打开新页面。

而如果使用 React 框架，则必须显式的调用 preventDefault()方法来实现阻止默认行为，具体代码如下：

【代码 5-4】

```
01  // TODO: React Event
02  function Hyperlink() {
03      function handleClick(e) {
04          e.preventDefault();
05          console.log('Prevent hyperlink address open new page!');
06      }
07      return (
08          <a href="#" onClick={handleClick}>
```

```
09          Click Me!
10       </a>
11    );
12 }
```

关于【代码 5-4】的说明：

- 第 03~06 行代码是事件处理方法（handleClick）的实现过程，其中第 04 行代码通过参数 "e" 显式的调用 preventDefault() 方法来阻止事件的默认行为。

另外，参数 "e" 是一个合成事件。React 框架根据 W3C 规范来定义这些合成事件，所以设计人员一般不需要担心浏览器的兼容性问题。

以上就是 React 框架中对于阻止事件默认行为的规定，与传统的 HTML DOM 事件处理方式还是不同的。

5.2 React 单击事件

首先，我们就从最简单的 React 单击事件（onClick）开始介绍，先使用最基本 JSX 方式实现一个单击事件的处理方法，具体代码如下：

【代码 5-5】（详见源代码目录 ch05-react-event-onClick.html 文件）

```
01 <!DOCTYPE html>
02 <html>
03 <head>
04     <meta charset="UTF-8"/>
05     <title>React Event - Click</title>
06     <script src="https://unpkg.com/react@16/umd/react.development.js"></script>
07     <script src="https://unpkg.com/react-dom@16/umd/react-dom.development.js"></script>
08     <!-- Don't use this in production: -->
09     <script src="https://unpkg.com/babel-standalone@6.15.0/babel.min.js"></script>
10 </head>
11 <body>
12 <!-- 添加文档主体内容 -->
13 <div id='id-div-react'></div>
14 <script type="text/babel">
15     // TODO: get div
16     var divReact = document.getElementById('id-div-react');
17     // TODO: function component
```

```
18      function onBtnClick() {
19          console.log("Clicked OK!");
20      }
21      // TODO: React JSX
22      const reactSpan = (
23          <span>
24              <h3>React Event - Basic Click</h3>
25              <button onClick={onBtnClick}>React Click</button>
26          </span>
27      );
28      // TODO: React render
29      ReactDOM.render(reactSpan, divReact);
30  </script>
31  </body>
32  </html>
```

关于【代码 5-5】的说明：

- 核心代码是第 25 行代码定义的按钮单击事件处理方法，单击事件名称为小驼峰写法为 "onClick"，事件处理的函数方法为 JSX 方式的 "{onBtnClick}"。
- 第 18～20 行代码是事件处理方法（onBtnClick）的实现过程。

测试网页的效果如图 5.1 所示。如图中的箭头所示，在页面中单击按钮后，浏览器控制台中输出了第 19 行代码定义的日志信息。

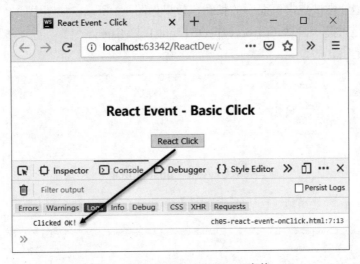

图 5.1　React 单击（onClick）事件

而如果采用传统的 HTML DOM 事件语法的书写方法会有什么结果呢？我们尝试将第 25 行代码改回传统的 HTML DOM 事件处理方式，具体代码如下：

【代码 5-6】（详见源代码目录 ch05-react-event-onClick.html 文件）

```
25        <button onclick="onBtnClick()">JavaScript Click</button>
```

测试网页，页面初始化后的效果如图 5.2 所示。如图中的箭头所示，页面在初始化后浏览器控制台就给出了错误提示信息（Warning: expected "onClick" listener…），说明事件方法名称（onclick）是错误的。

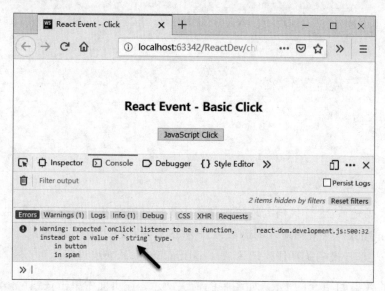

图 5.2　JavaScript 单击（onclick）事件（一）

然后，我们继续尝试点击页面中的按钮，效果如图 5.3 所示。如图中的箭头所示，单击按钮后继续给出了错误提示信息（Error: expected "onClick" listener…），进一步证明了事件方法名称（onclick）是错误的。

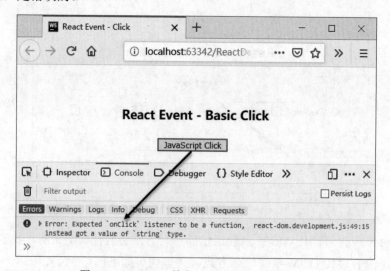

图 5.3　JavaScript 单击（onclick）事件（二）

5.3 React 阻止事件默认行为

在本节中,我们介绍 React 框架中如何使用 preventDefault()方法阻止事件的默认行为,具体代码如下:

【代码 5-7】(详见源代码目录 ch05-react-event-preventDefault.html 文件)

```
01   <!DOCTYPE html>
02   <html>
03   <head>
04       <meta charset="UTF-8"/>
05       <title>React Event - preventDefault</title>
06       <script src="https://unpkg.com/react@16/umd/react.development.js"></script>
07       <script src="https://unpkg.com/react-dom@16/umd/react-dom.development.js"></script>
08       <!-- Don't use this in production: -->
09       <script src="https://unpkg.com/babel-standalone@6.15.0/babel.min.js"></script>
10   </head>
11   <body>
12   <!-- 添加文档主体内容 -->
13   <div id='id-div-react'></div>
14   <script type="text/babel">
15       // TODO: get div
16       var divReact = document.getElementById('id-div-react');
17       // TODO: function component
18       function PreventLink() {
19           function handleClick(e) {
20               e.preventDefault();
21               console.log("The link has been clicked!");
22           }
23           return (
24               <a href="https://reactjs.org/" onClick={handleClick}>
25                   Click Me
26               </a>
27           );
28       }
```

```
29      // TODO: React JSX
30      const reactSpan = (
31          <span>
32              <h3>React Event - preventDefault</h3>
33              <PreventLink />
34          </span>
35      );
36      // TODO: React render
37      ReactDOM.render(reactSpan, divReact);
38  </script>
39  </body>
40  </html>
```

关于【代码 5-7】的说明：

- 第 24~26 行代码定义了一个超链接（<a>）元素。其中，"href"属性定义了一个网址（https://reactjs.org/）；同时还定义了单击事件（onClick）处理方法，事件方法名称为（handleClick）。
- 第 19~22 行代码是事件处理方法（handleClick）的实现过程，其中第 20 行代码通过合成事件参数（e）调用 preventDefault()方法，阻止超链接（<a>）单击事件的默认行为。

测试网页的效果如图 5.4 所示。如图中的箭头所示，在页面中单击超链接（Click Me）后，页面并没有跳转到预定义的网址（https://reactjs.org/）中，而浏览器控制台中输出了第 21 行代码定义的日志信息，说明 preventDefault()方法成功阻止了超链接单击事件的默认行为。

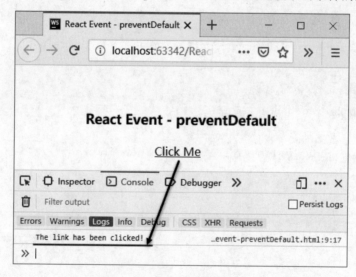

图 5.4　React 阻止事件默认行为

5.4 React 类的事件处理方法

在 React 框架中，如果使用 ES6 Class 语法来定义一个组件的时候，那么事件处理器会成为该类的一个方法。下面，我们看一个具体的代码实例：

【代码 5-8】（详见源代码目录 ch05-react-event-preventDefault.html 文件）

```
01  <!DOCTYPE html>
02  <html>
03  <head>
04      <meta charset="UTF-8"/>
05      <title>React Event - Class</title>
06      <script src="https://unpkg.com/react@16/umd/react.development.js"></script>
07      <script src="https://unpkg.com/react-dom@16/umd/react-dom.development.js"></script>
08      <!-- Don't use this in production: -->
09      <script src="https://unpkg.com/babel-standalone@6.15.0/babel.min.js"></script>
10  </head>
11  <body>
12  <!-- 添加文档主体内容 -->
13  <div id='id-div-react'></div>
14  <script type="text/babel">
15      // TODO: get div
16      var divReact = document.getElementById('id-div-react');
17      // TODO: define ES6 Class Component
18      class BtnClickComp extends React.Component {
19          constructor(props) {
20              super(props);
21              // TODO: 为了在回调方法中使用'this'，这个绑定是必不可少的
22              this.handleClick = this.handleClick.bind(this);
23          }
24          handleClick() {
25              console.log("React Event Class - Clicked OK!");
26          }
27          render() {
28              return (
29                  <button onClick={this.handleClick}>
30                      Click Me!
31                  </button>
32              );
33          }
34      }
35      // TODO: React JSX
```

```
36        const reactSpan = (
37            <span>
38                <h3>React Event - Class</h3>
39                <BtnClickComp />
40            </span>
41        );
42        // TODO: React render
43        ReactDOM.render(reactSpan, divReact);
44    </script>
45 </body>
46 </html>
```

关于【代码 5-8】的说明：

- 第 18～34 行代码定义了一个 ES6 类组件（BtnClickComp），下面会在这个类组件中实现一个事件处理方法。
- 第 29～31 行代码定义了一个按钮（<button>）元素，并定义了单击事件（onClick）处理方法，事件方法名称为（handleClick）。
- 第 24～26 行代码是事件处理方法（handleClick）的实现过程，第 25 行代码会在浏览器控制台中输出一行调试信息。
- 比较关键的代码是在第 24～26 行代码定义的 constructor()构造方法中，第 22 行代码通过 bind()方法为回调方法 "this.handleClick" 绑定了 this 关键字。其实，这点并不是 React 框架的特有行为，而是与 JavaScript 函数的工作原理有关。在 JavaScript 中，如果没有在方法名称后面加上小括号 "()"，那么必须为这个方法绑定上 this 关键字，否则调用方法时就会返回 "undefined" 错误。

测试网页的效果如图 5.5 所示。如图中的箭头所示，在页面中单击按钮（Click Me）后，浏览器控制台成功输出了第 25 行代码定义的调试信息。

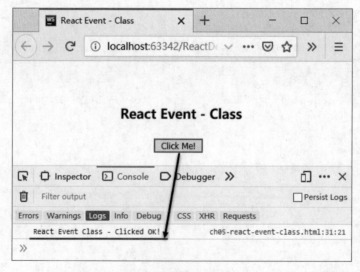

图 5.5　React 类的事件处理方法（一）

假如没有经过第 22 行代码所定义的、绑定 this 关键字的操作呢？那么在回调方法中使用 this 关键字时，就会报错提示 this 关键字为未定义（undefined）。具体效果如图 5.6 所示。如图中箭头所示，在页面中单击按钮（Click Me）后，浏览器控制台提示了错误信息（TypeError: this is undefined）。

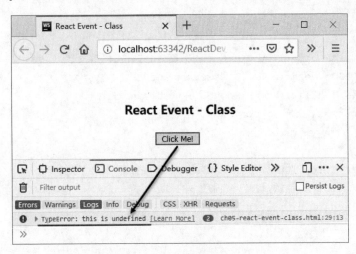

图 5.6　React 类的事件处理方法（二）

5.5　绑定回调方法的其他方式

在前一节中，我们介绍了如何为 React 类定义事件处理方法，还着重强调了绑定 this 关键字的重要性。但如果读者觉得这种通过 bind()方法进行绑定的操作比较烦琐，其实还有其他方式可以绑定事件处理的回调方法。

首先，就是通过实验性的 public class fields 语法正确地绑定回调方法的方式。请看下面的代码实例：

【代码 5-9】（详见源代码目录 ch05-react-event-class-field.html 文件）

```
01  <!DOCTYPE html>
02  <html>
03  <head>
04      <meta charset="UTF-8"/>
05      <title>React Event - Class Field</title>
06      <script src="https://unpkg.com/react@16/umd/react.development.js"></script>
07      <script src="https://unpkg.com/react-dom@16/umd/react-dom.development.js"></script>
08      <!-- Don't use this in production: -->
09      <script src="https://unpkg.com/babel-standalone@6.15.0/babel.min.js"></script>
```

```
10    </head>
11    </style>
12    <body>
13    <!-- 添加文档主体内容 -->
14    <div id='id-div-react'></div>
15    <script type="text/babel">
16        // TODO: get div
17        var divReact = document.getElementById('id-div-react');
18        // TODO: define ES6 Class Component
19        class BtnClickComp extends React.Component {
20            constructor(props) {
21                super(props);
22            }
23            // TODO: 此语法确保'handleClick'内的'this'已被绑定
24            // TODO: 注意,这是 *实验性*的语法
25            handleClick = () => {
26                console.log("Class Field 'this' :", this);
27            };
28            render() {
29                return (
30                    <button onClick={this.handleClick}>
31                        Click Me!
32                    </button>
33                );
34            }
35        }
36        // TODO: React JSX
37        const reactSpan = (
38          <span>
39              <h3>React Event - Class Field</h3>
40              <BtnClickComp />
41          </span>
42        );
43        // TODO: React render
44        ReactDOM.render(reactSpan, divReact);
45    </script>
46    </body>
47    </html>
```

关于【代码 5-9】的说明：

- 关键代码是第 25~27 行代码定义的回调方法（handleClick），这里是通过箭头函数方式定义的，这会确保 this 关键字被绑定。另外，根据 React 官方文档的说明，该绑定方式是"实验性"的语法。

测试网页的效果如图 5.7 所示。如图中的箭头所示，在页面中单击按钮（Click Me）后，浏览器控制台成功输出了关于 this 关键字的调试信息，说明成功的绑定了 this 关键字。

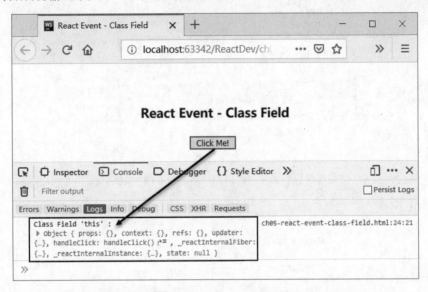

图 5.7　React 事件 Class Field 语法绑定 this

除了实验性的 public class fields 语法，还可以在事件方法的定义中，直接在回调方法中使用箭头函数的方式实现绑定 this 关键字。请看下面的代码实例：

【代码 5-10】（详见源代码目录 ch05-react-event-class-field.html 文件）

```
01  <!DOCTYPE html>
02  <html>
03  <head>
04      <meta charset="UTF-8"/>
05      <title>React Event - Arrow Func</title>
06      <script src="https://unpkg.com/react@16/umd/react.development.js"></script>
07      <script src="https://unpkg.com/react-dom@16/umd/react-dom.development.js"></script>
08      <!-- Don't use this in production: -->
09      <script src="https://unpkg.com/babel-standalone@6.15.0/babel.min.js"></script>
10  </head>
11  <body>
12  <!-- 添加文档主体内容 -->
13  <div id='id-div-react'></div>
14  <script type="text/babel">
15      // TODO: get div
```

```
16      var divReact = document.getElementById('id-div-react');
17      // TODO: define ES6 Class Component
18      class BtnClickComp extends React.Component {
19          constructor(props) {
20              super(props);
21          }
22          handleClick = () => {
23              console.log("Arrow func 'this' :", this);
24          };
25          render() {
26              // 此语法确保'handleClick'内的'this'已被绑定
27              return (
28                  <button onClick={(e) => this.handleClick(e)}>
29                      Click Me!
30                  </button>
31              );
32          }
33      }
34      // TODO: React JSX
35      const reactSpan = (
36          <span>
37              <h3>React Component - Arrow Func</h3>
38              <BtnClickComp />
39          </span>
40      );
41      // TODO: React render
42      ReactDOM.render(reactSpan, divReact);
43  </script>
44  </body>
45  </html>
```

关于【代码5-10】的说明：

- 关键代码是第 28 行代码为单击事件（onClick）定义的箭头函数，这里是通过箭头函数方式绑定 this 关键字的。

测试网页的效果如图 5.8 所示。如图中的箭头所示，在页面中单击按钮（Click Me）后，浏览器控制台成功输出了关于 this 关键字的调试信息，说明 this 关键字也被成功绑定了。

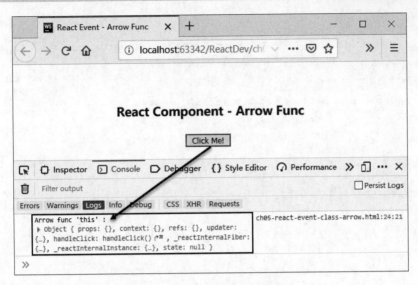

图 5.8　React 事件箭头函数绑定 this

读者注意，【代码 5-10】方式的问题在于每次渲染组件时都会创建不同的回调函数。多数情况下不会有问题，但如果该回调函数作为参数 Props 传入子组件时，这些组件可能就会进行额外的重新渲染。因此，还是建议设计人员在构造器中绑定或使用 Class Fields 语法绑定这两种方式。

5.6　在事件处理方法中传递参数

前文中介绍了如何在 React 框架中实现事件处理方法，还介绍了几种绑定 this 关键字的方式。不多，读者应该会感觉到缺少了些重要的东西，如果仅仅是定义了事件处理方法是实现不了复杂功能的。

React 事件处理应该也可以像 JavaScript 程序那样，在事件处理方法中通过传递参数实现更复杂的功能。实际情况也确实如此，React 可以通过两种方式在事件处理方法中传递参数，分别是箭头函数方式和通过 bind()方法的绑定方式。

首先，我们看一下通过箭头函数传递参数的方法，请看下面的代码实例：

【代码 5-11】（详见源代码目录 ch05-react-event-params-arrow-func.html 文件）

```
01  <!DOCTYPE html>
02  <html>
03  <head>
04      <meta charset="UTF-8"/>
05      <title>React Event - Parameters</title>
06      <script src="https://unpkg.com/react@16/umd/react.development.js"></script>
```

```
07        <script src="https://unpkg.com/react-dom@16/umd/react-dom.development.js"></script>
08        <!-- Don't use this in production: -->
09        <script src="https://unpkg.com/babel-standalone@6.15.0/babel.min.js"></script>
10    </head>
11    <body>
12        <!-- 添加文档主体内容 -->
13        <div id='id-div-react'></div>
14        <script type="text/babel">
15            // TODO: get div
16            var divReact = document.getElementById('id-div-react');
17            // TODO: define ES6 Class Component
18            class ParamsEventComp extends React.Component{
19                constructor(props){
20                    super(props);
21                    this.state = {
22                        name: 'Hello React!'
23                    };
24                }
25                passParamsClick(name, e) {    // TODO: 事件对象 e 要放在最后
26                    e.preventDefault();
27                    console.log(name);
28                }
29                render(){
30                    return (
31                        <div>
32                            {/* 通过 Arrow Function 方法传递参数 */}
33                            <button onClick={(e) => this.passParamsClick(this.state.name, e)}>
34                                Click Me
35                            </button>
36                        </div>
37                    );
38                }
39            }
40            // TODO: React JSX
41            const reactSpan = (
42                <span>
43                    <h3>React Event - Parameters Arrow Func</h3>
44                    <ParamsEventComp />
45                </span>
46            );
```

```
47        // TODO: React render
48        ReactDOM.render(reactSpan, divReact);
49    </script>
50    </body>
51    </html>
```

关于【代码 5-11】的说明：

- 第 21～23 行代码在 ES6 类组件（ParamsEventComp）的 constructor()构造方法中，定义了一个 State 状态属性（name），并初始化了属性值。
- 第 33 行代码在按钮单击事件（onClick）定义中，通过箭头函数的方式定义了回调方法（passParamsClick），并传递了两个参数。
- 第 1 个参数是 State 状态属性（name）。
- 第 2 个参数是合成事件参数对象（e）。
- 第 25～28 行代码是事件处理方法（passParamsClick）的实现过程，注意要将合成事件参数对象（e）放在最后。

测试网页的效果如图 5.9 所示。如图中的箭头所示，在页面中单击按钮（Click Me）后，浏览器控制台成功输出了第 27 行代码定义的调试信息，说明参数传递的操作成功了。

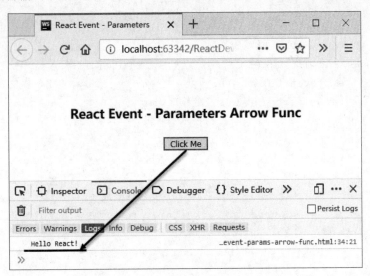

图 5.9　React 事件传递参数（一）

另外一种方式就是通过 bind()方法进行绑定来传递参数，请继续看下面的代码实例：

【代码 5-12】（详见源代码目录 ch05-react-event-params-bind.html 文件）

```
01    <!DOCTYPE html>
02    <html>
03    <head>
04        <meta charset="UTF-8"/>
05        <title>React Event - Arrow Func Pass Arguments</title>
```

```html
06        <script src="https://unpkg.com/react@16/umd/react.development.js"></script>
07        <script src="https://unpkg.com/react-dom@16/umd/react-dom.development.js"></script>
08        <!-- Don't use this in production: -->
09        <script src="https://unpkg.com/babel-standalone@6.15.0/babel.min.js"></script>
10    </head>
11    <body>
12    <!-- 添加文档主体内容 -->
13    <div id='id-div-react'></div>
14    <script type="text/babel">
15        // TODO: get div
16        var divReact = document.getElementById('id-div-react');
17        // TODO: define ES6 Class Component
18        class ParamsEventComp extends React.Component {
19            constructor(props) {
20                super(props);
21                this.state = {
22                    name: 'Hello React!'
23                };
24            }
25            passParamsClick(name, e) {    // TODO: 事件对象 e 要放在最后
26                e.preventDefault();
27                console.log(name);
28            }
29            render() {
30                return (
31                    <div>
32                        {/* 通过 bind() 方法绑定传递参数 */}
33                        <button onClick={this.preventPop.bind(this, this.state.name)}>
34                            Click Me
35                        </button>
36                    </div>
37                );
38            }
39        }
40        // TODO: React JSX
41        const reactSpan = (
42            <span>
43                <h3>React Event - Parameters bind()</h3>
44                <ParamsEventComp />
```

```
45            </span>
46        );
47        // TODO: React render
48        ReactDOM.render(reactSpan, divReact);
49    </script>
50 </body>
51 </html>
```

关于【代码 5-12】的说明：

- 第 33 行代码在按钮单击事件（onClick）回调方法的定义中，通过 bind()绑定的方式传递了两个参数。
- 第 1 个参数是关键字 this。
- 第 2 个参数是 State 状态属性（name）。

测试网页的效果如图 5.10 所示。如图中的箭头所示，在页面中单击按钮（Click Me）后，浏览器控制台成功输出了第 27 行代码定义的调试信息，说明参数传递的操作同样成功了。

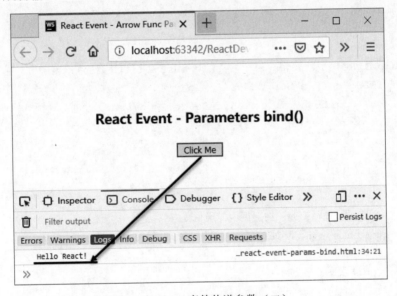

图 5.10　React 事件传递参数（二）

5.7　实战：开关按钮

本节在前面所介绍知识点的基础上，实现一个具有开关状态的按钮控件。通过这个"开关按钮"的实现过程，让读者体会一下 React 框架的强大功能，具体代码如下：

【代码 5-13】（详见源代码目录 ch05-react-event-toggle-btn.html 文件）

```html
01    <!DOCTYPE html>
02    <html>
03    <head>
04        <meta charset="UTF-8"/>
05        <title>React Event - Class</title>
06        <script src="https://unpkg.com/react@16/umd/react.development.js"></script>
07        <script src="https://unpkg.com/react-dom@16/umd/react-dom.development.js"></script>
08        <!-- Don't use this in production: -->
09        <script src="https://unpkg.com/babel-standalone@6.15.0/babel.min.js"></script>
10    </head>
11    <body>
12    <!-- 添加文档主体内容 -->
13    <div id='id-div-react'></div>
14    <script type="text/babel">
15        // TODO: get div
16        var divReact = document.getElementById('id-div-react');
17        // TODO: define ES6 Class Component
18        class ToggleBtnComp extends React.Component {
19            constructor(props) {
20                super(props);
21                this.state = {
22                    isToggleOn: true
23                };
24                // TODO: 为了在回调中使用'this'，这个绑定是必不可少的
25                this.handleClick = this.toggleBtnClick.bind(this);
26            }
27            toggleBtnClick(isToggleOn, e) {
28                e.preventDefault();
29                console.log("Now state is '" + isToggleOn + "' before to convert.");
30                this.setState(
31                    state => ({
32                        isToggleOn: !state.isToggleOn
33                    })
34                );
35            }
36            render() {
37                console.log("Now state is '" + this.state.isToggleOn + "'.");
38                return (
```

```
39                    <button onClick={(e) =>
this.toggleBtnClick(this.state.isToggleOn, e)}>
40                        {this.state.isToggleOn ? 'Button ON' : 'Button OFF'}
41                    </button>
42                );
43            }
44        }
45        // TODO: React JSX
46        const reactSpan = (
47            <span>
48                <h3>React Event - Toggle Button</h3>
49                <ToggleBtnComp />
50            </span>
51        );
52        // TODO: React render
53        ReactDOM.render(reactSpan, divReact);
54    </script>
55    </body>
56    </html>
```

关于【代码 5-13】的说明：

- 第 18～44 行代码定义了一个 ES6 类组件（ToggleBtnComp），用于实现一个"开关按钮"，详细说明如下：
  - 第 19～26 行代码定义的 constructor()构造方法中，定义了一个 State 状态属性（isToggleOn）用于标记"开关按钮"的状态，并初始化属性值为"true"。
  - 第 39～41 行代码在按钮单击事件（onClick）处理方法（toggleBtnClick）的定义中，通过箭头函数的方式定义了回调方法（toggleBtnClick），并传递了 State 状态属性（isToggleOn）和合成事件参数对象（e）这两个参数。其中，第 40 行代码通过三元条件表达式实现了"开关按钮"的状态切换。
  - 第 27～35 行代码是事件处理方法（toggleBtnClick）的实现过程。其中，第 30～34 行代码通过 setState()方法对 State 状态属性（isToggleOn）进行了重新赋值，也就是在"true"和"false"之间进行切换。
- 整个而言，ES6 类组件（ToggleBtnComp）通过对 State 状态属性（isToggleOn）值的切换，借助第 40 行代码实现了"开关按钮"的状态切换。具体表现就是按钮在每次点击后，按钮文本在"Button ON"和"Button OFF"之间进行切换。

测试网页，初始效果如图 5.11 所示。如图中箭头所示，页面初始状态下的按钮文本（Buuton ON）与浏览器控制台输出的调试信息（见第 37 行代码）相一致。

图 5.11　React 开关按钮（一）

然后，我们在页面中点击一次按钮，页面效果如图 5.12 所示。如图中的箭头所示，在点击一次按钮的操作后，浏览器控制台中先输出了切换前的 State 状态属性（isToggleOn）值（见第 29 行代码）。然后，页面中的按钮文本由（Buuton ON）切换为（Buuton OFF），且与浏览器控制台再次输出的调试信息（见第 37 行代码）相一致。

图 5.12　React 开关按钮（二）

5.8 React 文本框事件

前文中主要介绍了最基本的 React 单击事件（onClick）实例，读者可能还比较关心其他事件如何使用。在本节中，我们就详细介绍关于文本框（<input>）元素的几个常用事件的使用方法。

首先，就是关于文本输入框获取焦点事件（onFocus）的使用。下面看一下具体的代码实例：

【代码 5-14】（详见源代码目录 ch05-react-event-input-onFocus.html 文件）

```
01  class InputFocusComp extends React.Component {
02      constructor(props) {
03          super(props);
04          // TODO: 为了在回调中使用'this'，这个绑定是必不可少的
05          this.inputTextFocus = this.inputTextFocus.bind(this);
06      }
07      inputTextFocus(e) {
08          e.preventDefault();
09      }
10      render() {
11          return (
12              <input type="text" onFocus={this.inputTextFocus} />
13          );
14      }
15  }
```

关于【代码 5-14】的说明：

- 第 12 行代码定义了一个文本输入框（<input type="text">），并定义了获取焦点事件（onFocus）的回调方法"{inputTextFocus}"。
- 第 05 行代码通过 bind() 方法为回调方法"{inputTextFocus}"绑定了 this 关键字。
- 第 07~09 行代码是获取焦点事件（onFocus）回调方法（inputTextFocus）的实现过程。

接着，是关于文本输入框失去焦点事件（onBlur）的使用。下面看一下具体的代码实例：

【代码 5-15】（详见源代码目录 ch05-react-event-input-onBlur.html 文件）

```
01  class InputBlurComp extends React.Component {
02      constructor(props) {
03          super(props);
04          // TODO: 为了在回调中使用'this'，这个绑定是必不可少的
05          this.inputTextBlur = this.inputTextBlur.bind(this);
06      }
07      inputTextBlur(e) {
08          e.preventDefault();
```

```
09      }
10      render() {
11        return (
12          <input type="text" onBlur={this.inputTextBlur} />
13        );
14      }
15    }
```

关于【代码 5-15】的说明：

- 该段代码与【代码 5-14】类似，不同之处是失去焦点事件的名称为 onBlur。

最后是关于文本输入框中内容发生改变的事件（onChange）的使用。下面看一下具体的代码实例：

【代码 5-16】（详见源代码目录 ch05-react-event-input-onChange.html 文件）

```
01  class InputChangeComp extends React.Component {
02    constructor(props) {
03      super(props);
04      this.state = {
05        inputVal: ""
06      };
07      // TODO: 为了在回调中使用'this'，这个绑定是必不可少的
08      this.inputValueChange = this.inputValueChange.bind(this);
09    }
10    inputValueChange(e) {
11      e.preventDefault();
12      this.setState({
13        inputVal: e.target.value
14      });
15    }
16    render() {
17      const inputVal = this.state.inputVal;
18      return (
19        <input type="text"
20            value={inputVal}
21            onChange={this.inputValueChange} />
22      );
23    }
24  }
```

关于【代码 5-16】的说明：

- 在 React 框架中，文本输入框的 "onChange" 事件与 "onBlur" 和 "onFocus" 使用方法类似，不同之处是随着 "onChange" 事件的触发，大多数情况下需要监控文本框

中输入内容的变化。因此，需要增加如下的代码:
- 第 04～06 行代码定义了 State 状态属性（inputVal），用于同步文本输入框的内容。
- 第 10～15 行代码定义的"onChange"事件的回调方法中，通过 setState()方法更新 State 状态属性（inputVal）值，注意是通过合成事件对象（e）获取的文本输入框的内容。
- 第 20 行代码定义文本输入框的"value"属性，用于实时表现内容的变化。另外，React 框架还增加了一个"defaultValue"属性（只读不可更改），用于显示文本输入框的初始内容。需要注意的是，React 框架不支持同时使用"value"属性和"defaultValue"属性。

以上详细介绍了在 React 框架下，文本输入框（<input>）中最常用的"onFocus"事件、"onBlur"事件和"onChange"事件的使用方法。下面再介绍一个使用这三个文本输入框事件处理方法的代码实例：

【代码 5-17】（详见源代码目录 ch05-react-event-input-listener.html 文件）

```
01  <!DOCTYPE html>
02  <html>
03  <head>
04      <meta charset="UTF-8"/>
05      <title>React Event - Input Listener</title>
06      <script src="https://unpkg.com/react@16/umd/react.development.js"></script>
07      <script src="https://unpkg.com/react-dom@16/umd/react-dom.development.js"></script>
08      <!-- Don't use this in production: -->
09      <script src="https://unpkg.com/babel-standalone@6.15.0/babel.min.js"></script>
10  </head>
11  <body>
12  <!-- 添加文档主体内容 -->
13  <div id='id-div-react'></div>
14  <script type="text/babel">
15      // TODO: get div
16      var divReact = document.getElementById('id-div-react');
17      // TODO: define ES6 Class Component
18      class InputListenerComp extends React.Component {
19          constructor(props) {
20              super(props);
21              this.state = {
22                  inputVal: ""
23              };
24              // TODO: 为了在回调中使用'this'，这个绑定是必不可少的
```

```
25              this.inputTextFocus = this.inputTextFocus.bind(this);
26              this.inputTextChange = this.inputTextChange.bind(this);
27              this.inputTextBlur = this.inputTextBlur.bind(this);
28          }
29          inputTextFocus(e) {
30              e.preventDefault();
31              console.log("input text focus.");
32              this.setState({
33                  inputVal: e.target.value
34              });
35          }
36          inputTextChange(e) {
37              e.preventDefault();
38              console.log("input text changed to '" + this.state.inputVal + "'.");
39              this.setState({
40                  inputVal: e.target.value
41              });
42          }
43          inputTextBlur(e) {
44              e.preventDefault();
45              console.log("input text blur.");
46              this.setState({
47                  inputVal: e.target.value
48              });
49          }
50          render() {
51              const inputVal = this.state.inputVal;
52              console.log("Now inputVal is '" + this.state.inputVal + "'.");
53              return (
54                  <input type="text"
55                      value={inputVal}
56                      onFocus={this.inputTextFocus}
57                      onChange={this.inputTextChange}
58                      onBlur={this.inputTextBlur} />
59              );
60          }
61      }
62      // TODO: React JSX
63      const reactSpan = (
64          <span>
65              <h3>React Event - Input Listener</h3>
66              <InputListenerComp />
```

```
67              </span>
68          );
69          // TODO: React render
70          ReactDOM.render(reactSpan, divReact);
71      </script>
72  </body>
73  </html>
```

关于【代码 5-17】的说明：

- 这段代码将文本输入框的 "onFocus" 事件、"onBlur" 事件和 "onChange" 事件全部进行了监控，在每次触发这三个事件时均向浏览器控制台输出了对应的调试信息。

测试网页的效果如图 5.13 所示。在页面上的文本输入框中依次输入字符 "abc" 后，浏览器控制台中输出了相应的调试信息。

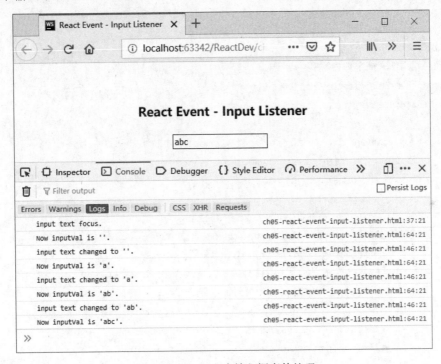

图 5.13　React 文本输入框事件处理

5.9　实战：水温监控控件

本节实现一个可以监控水温的文本控件，具体就是通过监控文本框中所输入的水温值，自动判断出该水温值的级别（本例中定义了冰水、冷水、温水、热水和沸水这几个级别），具体代码如下：

【代码 5-18】（详见源代码目录 ch05-react-event-water-temp.html 文件）

```
01  <!DOCTYPE html>
02  <html>
03  <head>
04      <meta charset="UTF-8"/>
05      <title>React Event - Water Temperature</title>
06      <script src="https://unpkg.com/react@16/umd/react.development.js"></script>
07      <script src="https://unpkg.com/react-dom@16/umd/react-dom.development.js"></script>
08      <!-- Don't use this in production: -->
09      <script src="https://unpkg.com/babel-standalone@6.15.0/babel.min.js"></script>
10  </head>
11  <body>
12  <!-- 添加文档主体内容 -->
13  <div id='id-div-react'></div>
14  <script type="text/babel">
15      // TODO: get div
16      var divReact = document.getElementById('id-div-react');
17      // TODO: define func Component
18      function WaterTempLevel(props) {
19          if (props.wlevel <= 0) {
20              return <p>This is ice water.</p>;
21          } else if((props.wlevel > 0) && (props.wlevel <= 20)) {
22              return <p>This is cold water.</p>;
23          } else if((props.wlevel > 20) && (props.wlevel <= 38)) {
24              return <p>This is warm water.</p>;
25          } else if((props.wlevel > 38) && (props.wlevel < 100)) {
26              return <p>This is hot water.</p>;
27          } else if(props.wlevel >= 100) {
28              return <p>This is boiling water.</p>;
29          } else {
30              return <p>This is ... water.</p>;
31          }
32      }
33      // TODO: define ES6 Class Component
34      class WaterTempComp extends React.Component {
35          constructor(props) {
36              super(props);
37              this.state = {
```

```
38              temperature: ""
39          };
40          // TODO: 为了在回调中使用'this', 这个绑定是必不可少的
41          this.inputTextFocus = this.inputTextFocus.bind(this);
42          this.inputTextChange = this.inputTextChange.bind(this);
43          this.inputTextBlur = this.inputTextBlur.bind(this);
44      }
45      inputTextFocus(e) {
46          e.preventDefault();
47          console.log("input text focus.");
48          this.setState({
49              temperature: e.target.value
50          });
51      }
52      inputTextChange(e) {
53          e.preventDefault();
54          console.log("input text changed.");
55          this.setState({
56              temperature: e.target.value
57          });
58      }
59      inputTextBlur(e) {
60          e.preventDefault();
61          console.log("input text blur.");
62          this.setState({
63              temperature: e.target.value
64          });
65      }
66      render() {
67          const temperature = this.state.temperature;
68          return (
69              <fieldset>
70                  <legend>Enter temperature to test:</legend>
71                  <input type="text"
72                      value={temperature}
73                      onFocus={this.inputTextFocus}
74                      onChange={this.inputTextChange}
75                      onBlur={this.inputTextBlur} />
76                  <WaterTempLevel wlevel={parseFloat(temperature)} />
77              </fieldset>
78          );
79      }
80  }
```

```
81          // TODO: React JSX
82          const reactSpan = (
83              <span>
84                  <h3>React Event - Water Temperature</h3>
85                  <WaterTempComp />
86              </span>
87          );
88          // TODO: React render
89          ReactDOM.render(reactSpan, divReact);
90      </script>
91  </body>
92  </html>
```

关于【代码5-18】的说明：

- 第34～80行代码定义了一个ES6类组件（WaterTempComp），用于实现一个"水温监控控件"，具体说明如下：
  - 第37～39行代码定义的constructor()构造方法中，定义了一个State状态属性（temperature）用于标记"水温监控控件"的温度值。
  - 第73～75行代码分别定义了文本输入框"onFocus"事件、"onChange"事件和"onBlur"事件的处理方法，第45～51行、第52～58行和第59～65行代码分别是这三个事件回调方法的实现过程。
  - 第76行代码引入了一个自定义组件（WaterTempLevel），并定义了一个Props参数（wlevel），用于传递用户输入的水温值。
- 第18～32行代码是自定义组件（WaterTempLevel）的实现过程，其中Props参数（wlevel）用于传递用户输入的水温值，然后第19～31行代码通过if条件语句判断温度值并返回水温级别。

测试网页的效果如图5.14至图5.17所示。

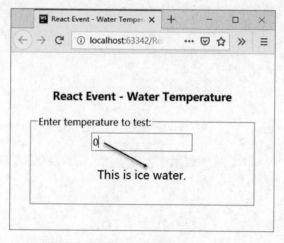

图5.14 React水温监控控件（一）

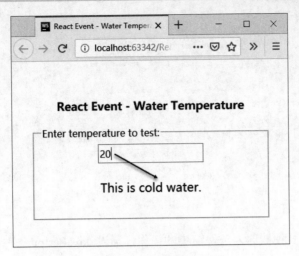

图 5.15　React 水温监控控件（二）

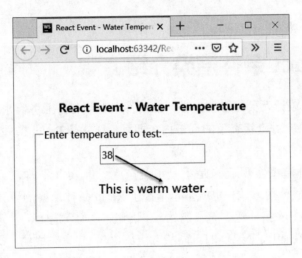

图 5.16　React 水温监控控件（三）

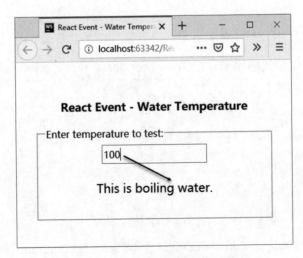

图 5.17　React 水温监控控件（四）

第 6 章

◀ React 条件渲染 ▶

React 框架还支持一种被称为条件渲染的操作,通过 JavaScript 条件运算符实现仅渲染满足条件部分的内容。

6.1 React 条件渲染介绍

在 React 框架中,设计人员可以创建不同的组件来封装各种业务需求,然后依据需求的不同状态,仅仅渲染组件中对应状态下的局部内容即可。这就是 React 条件渲染的存在逻辑,是一种真实存在的业务需求。

那么,React 条件渲染实现起来是不是很复杂呢?其实并不难,设计人员如果掌握了 JavaScript 条件运算符的使用方法,也就基本掌握了 React 条件渲染的设计方式。React 条件渲染基本上是通过 JavaScript 条件运算符(例如,if 语句、"与"逻辑、三元条件表达式,等等)创建元素组件,并让 React 根据组件的不同状态渲染更新 UI 来完成的。

下面,我们通过一个简单的表示用户当前登录状态的代码实例,介绍 React 条件渲染的基本使用方法。

(1)先创建两个组件,分别表示用户"已登录"和"未登录"的状态,具体代码如下:

【代码 6-1】(详见源代码目录 ch06-react-render-if-user-logged.html 文件)

```
01  function UserLoggedInComp(props) {
02      return <p>Hello, welcome back to my app.</p>;
03  }
04  function UserLoggedOutComp(props) {
05      return <p>Hello, please log in first.</p>;
06  }
```

关于【代码 6-1】的说明:

- 第 01~03 行代码定义了第一个函数组件(UserLoggedInComp),用于表示用户"已登录"的状态。

- 第 04~06 行代码定义了第二个函数组件（UserLoggedOutComp），用于表示用户"未登录"的状态。

（2）再创建一个组件，它将根据当前用户的登录状态，选择是显示用户"已登录"的组件（UserLoggedInComp）还是显示用户"未登录"的组件（UserLoggedOutComp），具体代码如下：

【代码 6-2】（详见源代码目录 ch06-react-render-if-user-logged.html 文件）

```
01  function UserLoggedComp(props) {
02      const isLogged = props.isLogged;
03      if(isLogged) {
04          return <UserLoggedInComp />;
05      } else {
06          return <UserLoggedOutComp />;
07      }
08  }
```

关于【代码 6-2】的说明：

- 第 02 行代码定义了一个 Props 参数（isLogged），用于表示用户当前的登录状态（true 表示"已登录"、false 表示"未登录"）。
- 第 03~07 行代码通过 if 条件选择语句，通过判断 Props 参数（isLogged）选择显示用户"已登录"的组件（UserLoggedInComp）或是用户"未登录"的组件（UserLoggedOutComp）。

（3）将【代码 6-1】和【代码 6-2】中的组件整合到一起，通过 Props 参数（isLogged）传递当前用户的登录状态，具体如下：

【代码 6-3】（详见源代码目录 ch06-react-render-if-user-logged.html 文件）

```
01  <!DOCTYPE html>
02  <html>
03  <head>
04      <meta charset="UTF-8"/>
05      <title>React Render - if</title>
06      <script src="https://unpkg.com/react@16/umd/react.development.js"></script>
07      <script src="https://unpkg.com/react-dom@16/umd/react-dom.development.js"></script>
08      <!-- Don't use this in production: -->
09      <script src="https://unpkg.com/babel-standalone@6.15.0/babel.min.js"></script>
10  </head>
11  <body>
```

```
12      <!-- 添加文档主体内容 -->
13      <div id='id-div-react'></div>
14      <script type="text/babel">
15          // TODO: get div
16          var divReact = document.getElementById('id-div-react');
17          // TODO: function component
18          function UserLoggedInComp(props) {
19              return <p>Hello, welcome back to my app.</p>;
20          }
21          function UserLoggedOutComp(props) {
22              return <p>Hello, please log in first.</p>;
23          }
24          // TODO: function component - conditional render
25          function UserLoggedComp(props) {
26              const isLogged = props.isLogged;
27              if(isLogged) {
28                  return <UserLoggedInComp />;
29              } else {
30                  return <UserLoggedOutComp />;
31              }
32          }
33          // TODO: React JSX
34          const reactSpan = (
35            <span>
36                <h3>React Render - If User Logged In?</h3>
37                <UserLoggedComp isLogged={true} />
38            </span>
39          );
40          // TODO: React render
41          ReactDOM.render(reactSpan, divReact);
42      </script>
43      </body>
44      </html>
```

关于【代码 6-3】的说明：

- 在第 37 行代码所引入的元素组件 \<UserLoggedComp\> 中，添加了一个属性（isLogged），并初始化为 "true"（true 表示 "已登录"）。该属性值将作为 Props 参数传递到第 25~32 行代码定义的组件（UserLoggedComp）中，用于表示用户当前的登录状态。

测试网页的效果如图 6.1 所示。如图中的箭头所示，页面中显示的信息表示用户当前是 "已登录" 状态。

图 6.1　React 条件渲染——用户登录状态（一）

然后，再尝试将【代码 6-3】中第 37 行代码中的布尔值改为 "false"，重新使用 Firefox 浏览器运行测试该 HTML 网页，具体效果如图 6.2 所示。如图中的箭头所示，页面中显示的信息表示用户当前是 "未登录" 状态。

图 6.2　React 条件渲染——用户登录状态（二）

以上实例通过 if 条件语句实现的 React 条件渲染操作，可以看到 React 框架在使用中的灵活多样性。

6.2 元素变量的条件渲染

在 React 条件渲染中，设计人员还可以使用变量来储存元素，进行有条件地渲染组件中的一部分，而其他部分并不会因为该渲染的部分而改变。

下面通过一个条件渲染的用户 "登录" 按钮和 "注销" 按钮的应用，介绍元素变量的条件渲染过程。

（1）先创建两个函数组件（LoginButton 和 LogoutButton），分别用于表示 "登录" 按钮和 "注销" 按钮的元素变量，具体代码如下：

【代码 6-4】（详见源代码目录 ch06-react-render-login-logout.html 文件）

```
01    // TODO: function component - login button
02    function LoginButton(props) {
03        return (
04            <button onClick={props.onClick}>
05                Login
06            </button>
07        );
08    }
09    // TODO: function component - logout button
10    function LogoutButton(props) {
11        return (
12            <button onClick={props.onClick}>
13                Logout
14            </button>
15        );
16    }
```

关于【代码 6-4】的说明：

- 第 02～08 行代码定义了第一个函数组件（LoginButton），用于表示"登录"按钮，并通过参数 Props 定义了单击事件（onClick）处理方法。
- 第 10～16 行代码定义了第二个函数组件（LogoutButton），用于表示"注销"按钮，同样通过参数 Props 定义了单击事件（onClick）处理方法。

（2）再创建一个 ES6 类组件（LoginControl），将根据 State 状态属性判断选择是渲染"登录"按钮组件、还是"注销"按钮组件。关于 ES6 类组件（LoginControl）的具体代码如下：

【代码 6-5】（详见源代码目录 ch06-react-render-login-logout.html 文件）

```
01    // TODO: define ES6 Class Component
02    class LoginControl extends React.Component {
03        constructor(props) {
04            super(props);
05            this.handleLoginClick = this.handleLoginClick.bind(this);
06            this.handleLogoutClick = this.handleLogoutClick.bind(this);
07            this.state = {isLoggedIn: false};
08        }
09        handleLoginClick() {
10            this.setState({isLoggedIn: true});
11        }
12        handleLogoutClick() {
13            this.setState({isLoggedIn: false});
14        }
```

```
15      render() {
16          const isLoggedIn = this.state.isLoggedIn;
17          let button;
18          if (isLoggedIn) {
19              button = <LogoutButton onClick={this.handleLogoutClick} />;
20          } else {
21              button = <LoginButton onClick={this.handleLoginClick} />;
22          }
23          return (
24              <div>
25                  <UserLoggedComp isLoggedIn={isLoggedIn} />
26                  {button}
27              </div>
28          );
29      }
30  }
```

关于【代码 6-5】的说明：

- 第 07 行代码定义了一个 State 状态属性（isLoggedIn），并初始化为 "false"，用于表示登录组件（LoginControl）的状态。
    - isLoggedIn 属性值为 true，表示用户 "已登录"，页面会显示 "注销" 按钮。
    - isLoggedIn 属性值为 false，表示用户 "已注销"，页面会显示 "登录" 按钮。
- 第 15～29 行代码通过 render()方法进行页面组件的渲染操作，具体说明如下：
    - 第 16 行代码先获取了 State 状态属性（isLoggedIn）值。
    - 第 18～22 行代码通过 if 条件选择语句判断状态属性（isLoggedIn）的布尔值，根据判断结果选择显示 "登录" 按钮组件（LoginButton）或 "注销" 按钮组件（LogoutButton）。

（3）将【代码 6-4】和【代码 6-5】中的组件整合到一起，具体代码如下：

【代码 6-6】（详见源代码目录 ch06-react-render-login-logout.html 文件）

```
01  <!DOCTYPE html>
02  <html>
03  <head>
04      <meta charset="UTF-8"/>
05      <title>React Render - Element Variable</title>
06      <script src="https://unpkg.com/react@16/umd/react.development.js"></script>
07      <script src="https://unpkg.com/react-dom@16/umd/react-dom.development.js"></script>
08      <!-- Don't use this in production: -->
09      <script
```

```
         src="https://unpkg.com/babel-standalone@6.15.0/babel.min.js"></script>
      10    </head>
      11    </style>
      12    <body>
      13    <!-- 添加文档主体内容 -->
      14    <div id='id-div-react'></div>
      15    <script type="text/babel">
      16        // TODO: get div
      17        var divReact = document.getElementById('id-div-react');
      18        // TODO: function component
      19        function UserLoggedInComp(props) {
      20            return <p>Hello, welcome back to my app.</p>;
      21        }
      22        function UserLoggedOutComp(props) {
      23            return <p>Hello, please login first.</p>;
      24        }
      25        // TODO: function component - conditional render
      26        function UserLoggedComp(props) {
      27            const isLoggedIn = props.isLoggedIn;
      28            if(isLoggedIn) {
      29                return <UserLoggedInComp />;
      30            } else {
      31                return <UserLoggedOutComp />;
      32            }
      33        }
      34        // TODO: function component - login button
      35        function LoginButton(props) {
      36            return (
      37                <button onClick={props.onClick}>
      38                    Login
      39                </button>
      40            );
      41        }
      42        // TODO: function component - logout button
      43        function LogoutButton(props) {
      44            return (
      45                <button onClick={props.onClick}>
      46                    Logout
      47                </button>
      48            );
      49        }
      50        // TODO: define ES6 Class Component
      51        class LoginControl extends React.Component {
```

```
52        constructor(props) {
53            super(props);
54            this.handleLoginClick = this.handleLoginClick.bind(this);
55            this.handleLogoutClick = this.handleLogoutClick.bind(this);
56            this.state = {isLoggedIn: false};
57        }
58        handleLoginClick() {
59            this.setState({isLoggedIn: true});
60        }
61        handleLogoutClick() {
62            this.setState({isLoggedIn: false});
63        }
64        render() {
65            const isLoggedIn = this.state.isLoggedIn;
66            let button;
67            if (isLoggedIn) {
68                button = <LogoutButton onClick={this.handleLogoutClick} />;
69            } else {
70                button = <LoginButton onClick={this.handleLoginClick} />;
71            }
72            return (
73                <div>
74                    <UserLoggedComp isLoggedIn={isLoggedIn} />
75                    {button}
76                </div>
77            );
78        }
79    }
80    // TODO: React JSX
81    const reactSpan = (
82        <span>
83            <h3>React Render - Login & Logout</h3>
84            <LoginControl />
85        </span>
86    );
87    // TODO: React render
88    ReactDOM.render(reactSpan, divReact);
89 </script>
90 </body>
91 </html>
```

关于【代码6-6】的说明：

- 第74行代码调用的函数组件（UserLoggedComp）用于显示页面提示信息。

- 第 18~33 行代码是函数组件<UserLoggedComp>的实现过程，此处引用了【代码 6-3】的内容。

测试网页的效果如图 6.3 所示。如图中的箭头所示，页面中显示的信息表示用户当前是"已注销"状态，需要点击"Login"按钮进行登录操作。然后，尝试点击一下"Login"按钮，页面效果如图 6.4 所示。如图中的箭头所示，页面中显示的信息已经变更为"已登录"状态，标题按钮也切换为"Logout"字样了。

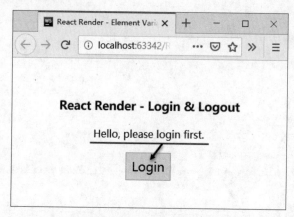

图 6.3　React 条件渲染——元素变量（一）

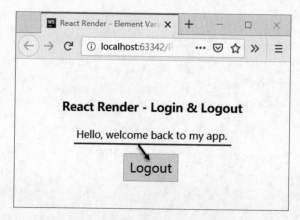

图 6.4　React 条件渲染——元素变量（二）

其实，【代码 6-6】还可以再进一步提炼以简化代码，将【代码 6-3】的内容提取出来整合到 ES6 类组件（LoginControl）中去，具体代码如下：

【代码 6-7】（详见源代码目录 ch06-react-render-btn-logged.html 文件）

```
01  <!DOCTYPE html>
02  <html>
03      <head>
04          <meta charset="UTF-8"/>
05          <title>React Render - Element Variable</title>
06          <script
```

```
       src="https://unpkg.com/react@16/umd/react.development.js"></script>
07         <script
       src="https://unpkg.com/react-dom@16/umd/react-dom.development.js"></script>
08         <!-- Don't use this in production: -->
09         <script
       src="https://unpkg.com/babel-standalone@6.15.0/babel.min.js"></script>
10     </head>
11     <body>
12     <!-- 添加文档主体内容 -->
13     <div id='id-div-react'></div>
14     <script type="text/babel">
15         // TODO: get div
16         var divReact = document.getElementById('id-div-react');
17         // TODO: function component - login button
18         function LoginButton(props) {
19             return (
20                 <fieldset>
21                     <p>Hello, please login first.</p>
22                     <button onClick={props.onClick}>
23                         Login
24                     </button>
25                 </fieldset>
26             );
27         }
28         // TODO: function component - logout button
29         function LogoutButton(props) {
30             return (
31                 <fieldset>
32                     <p>Hello, welcome back to my app.</p>
33                     <button onClick={props.onClick}>
34                         Logout
35                     </button>
36                 </fieldset>
37             );
38         }
39         // TODO: define ES6 Class Component
40         class LoginControl extends React.Component {
41             constructor(props) {
42                 super(props);
43                 this.handleLoginClick = this.handleLoginClick.bind(this);
44                 this.handleLogoutClick = this.handleLogoutClick.bind(this);
45                 this.state = {isLoggedIn: false};
46             }
```

```
47          handleLoginClick() {
48              this.setState({isLoggedIn: true});
49          }
50          handleLogoutClick() {
51              this.setState({isLoggedIn: false});
52          }
53          render() {
54              const isLoggedIn = this.state.isLoggedIn;
55              let button;
56              if (isLoggedIn) {
57                  button = <LogoutButton onClick={this.handleLogoutClick} />;
58              } else {
59                  button = <LoginButton onClick={this.handleLoginClick} />;
60              }
61              return (
62                  <div>
63                      {button}
64                  </div>
65              );
66          }
67      }
68      // TODO: React JSX
69      const reactSpan = (
70          <span>
71              <h3>React Render - Login & Logout</h3>
72              <LoginControl />
73          </span>
74      );
75      // TODO: React render
76      ReactDOM.render(reactSpan, divReact);
77  </script>
78  </body>
79  </html>
```

关于【代码6-7】的说明：

- 第 18~27 行代码定义了第一个函数组件（LoginButton），已经将【代码6-6】中定义的函数组件（UserLoggedComp）整合进去了。

同样的，第 29~38 行代码定义了第二个函数组件（LogoutButton），也将【代码6-6】中定义的函数组件（UserLoggedComp）整合进去了。

读者可自行使用 Firefox 浏览器运行测试该 HTML 网页，效果与图 6.3 和图 6.4 所示的基本是一样的。

6.3 逻辑"与"运算符的条件渲染

在 React 框架中,设计人员还可以借助 JSX 语法嵌入各种条件表达式,来实现条件渲染操作。例如,通过花括号将 JavaScript 逻辑与(&&)运算符表达式包裹进去,可以很方便地对元素组件进行"与"运算的条件渲染。

下面介绍一个通过逻辑与(&&)运算符表达式,实现一个判断测试成绩是否通过的应用,具体代码如下:

【代码 6-8】(详见源代码目录 ch06-react-render-and.html 文件)

```
01  <!DOCTYPE html>
02  <html>
03  <head>
04      <meta charset="UTF-8"/>
05      <title>React Render - AND(&&)</title>
06      <script src="https://unpkg.com/react@16/umd/react.development.js"></script>
07      <script src="https://unpkg.com/react-dom@16/umd/react-dom.development.js"></script>
08      <!-- Don't use this in production: -->
09      <script src="https://unpkg.com/babel-standalone@6.15.0/babel.min.js"></script>
10  </head>
11  <body>
12  <!-- 添加文档主体内容 -->
13  <div id='id-div-react'></div>
14  <script type="text/babel">
15      // TODO: get div
16      var divReact = document.getElementById('id-div-react');
17      // TODO: function component
18      function TestPass(props) {
19          const score = props.score;
20          return (
21              <div>
22                  {
23                      score >= 60 &&
24                      <p>
25                          Hello, your score <b>{score}</b> has passed the test.
26                      </p>
27                  }
28              </div>
29          );
```

```
30          }
31          // TODO: define const
32          const score = 90;
33          // TODO: React JSX
34          const reactSpan = (
35              <span>
36                  <h3>React Render - AND(&&)</h3>
37                  <TestPass score={score} />
38              </span>
39          );
40          // TODO: React render
41          ReactDOM.render(reactSpan, divReact);
42      </script>
43  </body>
44  </html>
```

关于【代码 6-8】的说明：

- 第 18～30 行代码定义了一个函数组件（TestPass），用于表示测试成绩是否通过（60 分及以上表示通过），具体说明如下：
  - 第 19 行代码定义了一个常量（score），通过 Props 参数获取了测试成绩。
  - 第 23～26 行代码通过逻辑与（&&）运算判断测试成绩是否通过，若通过，则渲染相应的提示文本。
- 第 32 行代码定义了一个常量（score），用于定义测试成绩。
- 第 37 行代码引用了函数组件（TestPass），并添加了属性（score）用于传递测试成绩（{score}）。

测试网页的效果如图 6.5 所示。如图中的箭头所示，页面中显示的信息表示测试成绩已经通过了。然后，尝试将第 32 行代码定义的成绩修改 50，页面效果如图 6.6 所示。页面中没有显示任何提示信息，表示测试成绩未通过。

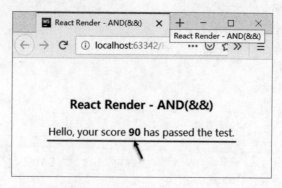

图 6.5　React 条件渲染——逻辑"与"运算（一）

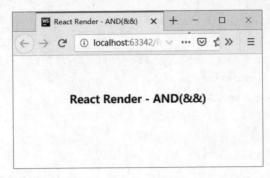

图 6.6　React 条件渲染——逻辑"与"运算（二）

为什么逻辑与（&&）运算可以以上的条件渲染效果呢？因为在 JavaScript 语法中，表达式（true && expression）总会返回 expression，而表达式（false && expression）则总会返回 false，因而可以实现 React 条件渲染的功能。

6.4　逻辑"或"运算符的条件渲染

在前一节的应用中，通过逻辑与（&&）运算符表达式实现了对测试成绩的判断操作。不过，相信读者会发现该代码有一定的不足，就是成绩未通过的情况下，页面中是无法显示提示信息的。

为了解决这个问题，我们通过引入逻辑或（||）运算符表达式，改进这个判断测试成绩是否通过的应用，具体代码如下：

【代码 6-9】（详见源代码目录 ch06-react-render-and-or.html 文件）

```
01  <!DOCTYPE html>
02  <html>
03  <head>
04      <meta charset="UTF-8"/>
05      <title>React Render - AND(&&)+OR(||)</title>
06      <script src="https://unpkg.com/react@16/umd/react.development.js"></script>
07      <script src="https://unpkg.com/react-dom@16/umd/react-dom.development.js"></script>
08      <!-- Don't use this in production: -->
09      <script src="https://unpkg.com/babel-standalone@6.15.0/babel.min.js"></script>
10  </head>
11  <body>
12  <!-- 添加文档主体内容 -->
13  <div id='id-div-react'></div>
```

```
14   <script type="text/babel">
15       // TODO: get div
16       var divReact = document.getElementById('id-div-react');
17       // TODO: function component
18       function TestPass(props) {
19           const score = props.score;
20           return (
21               <div>
22                   {
23                       (score >= 60 &&
24                       <p>
25                           Hello, your score <b>{score}</b> has passed the test.
26                       </p>)
27                       ||
28                       (score < 60 &&
29                       <p>
30                           Hello, your score <b>{score}</b> has not passed the test.
31                       </p>)
32                   }
33               </div>
34           );
35       }
36       // TODO: define const
37       const score = [90, 60, 50];
38       // TODO: React JSX
39       const reactSpan = (
40           <span>
41               <h3>React Render - AND(&&) + OR(||)</h3>
42               <TestPass score={score[0]} />
43               <TestPass score={score[1]} />
44               <TestPass score={score[2]} />
45           </span>
46       );
47       // TODO: React render
48       ReactDOM.render(reactSpan, divReact);
49   </script>
50   </body>
51   </html>
```

关于【代码 6-9】的说明：

- 第 18~35 行代码定义了一个函数组件（TestPass），用于表示测试成绩是否通过（60 分及以上表示通过），具体说明如下：

- 第 19 行代码定义了一个常量（score），通过 Props 参数获取了测试成绩；
- 第 22～32 行代码通过逻辑或（||）运算符将两个逻辑与（&&）运算符表达式（分别用于判断成绩通过与成绩没通过）并联起来，判断测试成绩是否通过，根据判断结果渲染所对应的提示文本。
- 第 32 行代码定义了一个常量（score）数组，定义了 3 个测试成绩。
● 第 37 行代码引用了函数组件（TestPass），并添加了属性（score）用于传递测试成绩（{score}）。

测试网页的效果如图 6.7 所示。如图中的箭头所示，页面中显示的信息已经可以区分测试成绩"通过"或"没通过"了。

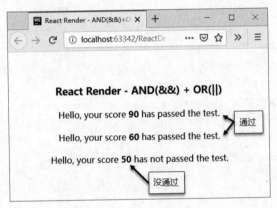

图 6.7　React 条件渲染——逻辑"或"运算

6.5 三元逻辑表达式的条件渲染

在 JavaScript 逻辑表达式中，还有一个比较常用的三元逻辑表达式，具体写法就是（condition ? true : false）。相信读者对该逻辑表达式已经很熟悉了，而且其与 React 条件渲染似乎更加契合。为什么会这么说呢？

我们还是通过具体的代码实例进行解释。在前一节的应用中，通过逻辑或（||）运算符表达式实现了对测试成绩"通过"与"没通过"的区分判断。不过【代码 6-9】中第 22～32 行代码的逻辑表达式看起来很烦琐，而且可读性也不好。于是，我们尝试用三元逻辑表达式改写这段条件渲染的操作，具体代码如下：

【代码 6-10】（详见源代码目录 ch06-react-render-and-or.html 文件）

```
01  <!DOCTYPE html>
02  <html>
03  <head>
04      <meta charset="UTF-8"/>
```

```
05      <title>React Render - Triple Condition</title>
06      <script src="https://unpkg.com/react@16/umd/react.development.js"></script>
07      <script src="https://unpkg.com/react-dom@16/umd/react-dom.development.js"></script>
08      <!-- Don't use this in production: -->
09      <script src="https://unpkg.com/babel-standalone@6.15.0/babel.min.js"></script>
10    </head>
11    <body>
12    <!-- 添加文档主体内容 -->
13    <div id='id-div-react'></div>
14    <script type="text/babel">
15      // TODO: get div
16      var divReact = document.getElementById('id-div-react');
17      // TODO: function component
18      function TestPass(props) {
19          const score = props.score;
20          return (
21            <div>
22              {
23                score >= 60 ?
24                <p>
25                    Hello, your score <b>{score}</b> has passed the test.
26                </p>
27                :
28                <p>
29                    Hello, your score <b>{score}</b> has not passed the test.
30                </p>
31              }
32            </div>
33          );
34      }
35      // TODO: define const
36      const score = [90, 60, 50];
37      // TODO: React JSX
38      const reactSpan = (
39        <span>
40            <h3>React Render - Triple Condition</h3>
41            <TestPass score={score[0]} />
42            <TestPass score={score[1]} />
43            <TestPass score={score[2]} />
```

```
44          </span>
45        );
46        // TODO: React render
47        ReactDOM.render(reactSpan, divReact);
48    </script>
49  </body>
50  </html>
```

关于【代码 6-10】的说明：

- 【代码 6-10】是在【代码 6-9】的基础上改写而成的，区别就是第 23～30 行代码所定义的、基于三元逻辑表达式的条件渲染操作。这段三元逻辑表达式通过判断测试成绩是否通过（≥60），来选择所对应的提示文本进行渲染。

测试网页的效果如图 6.8 所示。如图中的箭头所示，页面中显示的信息已经可以区分测试成绩"通过"或"没通过"了。

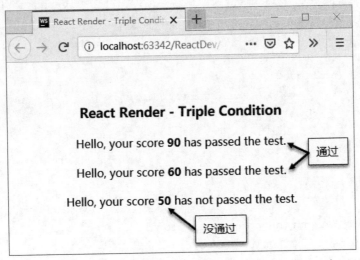

图 6.8　React 条件渲染——三元逻辑表达式

6.6　实战：改进登录组件

在前文的【代码 6-7】中，我们通过元素变量的条件渲染操作实现了一个登录组件。下面，我们借助三元逻辑表达式改进一下该登录组件的代码，具体代码如下：

【代码 6-11】（详见源代码目录 ch06-react-render-tri-condition-logged.html 文件）

```
01  <!DOCTYPE html>
02  <html>
03    <head>
```

```
04      <meta charset="UTF-8"/>
05      <title>React Render - Triple Condition</title>
06      <script src="https://unpkg.com/react@16/umd/react.development.js"></script>
07      <script src="https://unpkg.com/react-dom@16/umd/react-dom.development.js"></script>
08      <!-- Don't use this in production: -->
09      <script src="https://unpkg.com/babel-standalone@6.15.0/babel.min.js"></script>
10    </head>
11    <body>
12      <!-- 添加文档主体内容 -->
13      <div id='id-div-react'></div>
14      <script type="text/babel">
15        // TODO: get div
16        var divReact = document.getElementById('id-div-react');
17        // TODO: function component - login button
18        function LoginButton(props) {
19          return (
20            <fieldset>
21              <p>Hello, please login first.</p>
22              <button onClick={props.onClick}>
23                Login
24              </button>
25            </fieldset>
26          );
27        }
28        // TODO: function component - logout button
29        function LogoutButton(props) {
30          return (
31            <fieldset>
32              <p>Hello, welcome back to my app.</p>
33              <button onClick={props.onClick}>
34                Logout
35              </button>
36            </fieldset>
37          );
38        }
39        // TODO: define ES6 Class Component
40        class LoginControl extends React.Component {
41          constructor(props) {
42            super(props);
43            this.handleLoginClick = this.handleLoginClick.bind(this);
```

```
44          this.handleLogoutClick = this.handleLogoutClick.bind(this);
45          this.state = {isLoggedIn: false};
46      }
47      handleLoginClick() {
48          this.setState({isLoggedIn: true});
49      }
50      handleLogoutClick() {
51          this.setState({isLoggedIn: false});
52      }
53      render() {
54          const isLoggedIn = this.state.isLoggedIn;
55          return (
56              <div>
57                  {
58                      isLoggedIn ?
59                          <LogoutButton onClick={this.handleLogoutClick} />
60                          :
61                          <LoginButton onClick={this.handleLoginClick} />
62                  }
63              </div>
64          );
65      }
66  }
67  // TODO: React JSX
68  const reactSpan = (
69      <span>
70          <h3>React Render - Login & Logout</h3>
71          <LoginControl />
72      </span>
73  );
74  // TODO: React render
75  ReactDOM.render(reactSpan, divReact);
76  </script>
77  </body>
78  </html>
```

关于【代码 6-11】的说明：

- 【代码 6-11】是在【代码 6-7】的基础上改写而成的，区别就是第 57~62 行代码所定义的、基于三元逻辑表达式的条件渲染操作。这段三元逻辑表达式完全改写了原来的 if 条件选择表达式，通过判断 State 状态属性（isLoggedIn）的值，来选择所对应的按钮组件。

测试网页的效果如图 6.9 所示。如图中的箭头所示，通过点击页面中的按钮，可以切换"Login"和注销"Logout"状态。

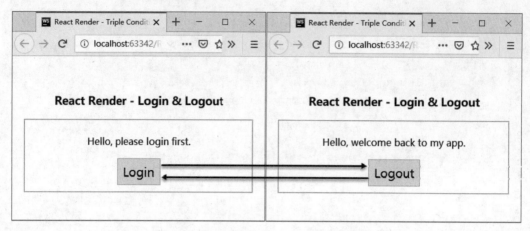

图6.9 React 条件渲染——改进登录组件

6.7 阻止组件渲染

在 React 开发的某些特殊场景下，设计人员可能需要隐藏特定的组件，即使这些组件已经被渲染完成了。如果要完成隐藏组件的操作，React 允许通过 render()方法直接返回空（null），进而达到不进行任何渲染的功能。

下面的代码实例中，我们定义一个函数组件（Banner）用于显示一个工具提示条。该函数组件会根据 Props 参数中的布尔值（isBanner）来进行条件渲染，如果布尔值（isBanner）为 false，就隐藏该组件（即不会渲染），具体代码如下：

【代码 6-12】（详见源代码目录 ch06-react-render-prevent-comp.html 文件）

```
01  <!DOCTYPE html>
02  <html>
03  <head>
04      <meta charset="UTF-8"/>
05      <title>React Render - Prevent Render</title>
06      <script src="https://unpkg.com/react@16/umd/react.development.js"></script>
07      <script src="https://unpkg.com/react-dom@16/umd/react-dom.development.js"></script>
08      <!-- Don't use this in production: -->
09      <script src="https://unpkg.com/babel-standalone@6.15.0/babel.min.js"></script>
10  </head>
```

```html
11  <body>
12    <!-- 添加文档主体内容 -->
13    <div id='id-div-react'></div>
14    <script type="text/babel">
15      // TODO: get div
16      var divReact = document.getElementById('id-div-react');
17      // TODO: function component
18      function Banner(props) {
19        if (!props.isBanner) {
20          return null;
21        }
22        return (
23          <div className="banner">
24            Banner div - show/hide by click the btn below.
25          </div>
26        );
27      }
28      // TODO: define ES6 Class Component
29      class BannerComp extends React.Component {
30        constructor(props) {
31          super(props);
32          this.state = {showBanner: true};
33          this.handleToggleClick = this.handleToggleClick.bind(this);
34        }
35        handleToggleClick() {
36          this.setState(prevState => ({
37            showBanner: !prevState.showBanner
38          }));
39        }
40        render() {
41          return (
42            <div>
43              <Banner isBanner={this.state.showBanner} />
44              <button onClick={this.handleToggleClick}>
45                {this.state.showBanner ? 'Hide' : 'Show'}
46              </button>
47            </div>
48          );
49        }
50      }
51      // TODO: React JSX
52      const reactSpan = (
53        <span>
```

```
54              <h3>React Render - Prevent Render</h3>
55              <BannerComp />
56          </span>
57      );
58      // TODO: React render
59      ReactDOM.render(reactSpan, divReact);
60 </script>
61 </body>
62 </html>
```

关于【代码6-12】的说明：

- 本代码核心部分就是第 18~27 行代码所定义的函数组件（Banner）。其中，第 19~21 行代码通过 if 条件语句判断 Props 参数（isBanner），如果为 false，就执行第 20 行代码直接返回空（null），从而达到隐藏组件的效果。
- 第 29~50 行代码定义了一个 ES6 类组件（BannerComp），它通过 State 状态属性（showBanner）的值切换按钮的状态。同时，在第 43 行代码中，通过将 State 状态属性（showBanner）的值传递给函数组件（Banner）的 Props 参数（isBanner），实现显示或隐藏组件（Banner）的效果。

测试网页的效果如图 6.10 所示。如图中的箭头所示，通过点击页面中的按钮（Hide | Show），可以实现工具提示条（Banner）"显示"与"隐藏"的切换效果。

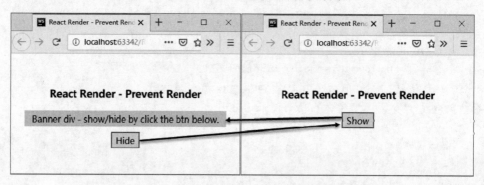

图 6.10　React 条件渲染——改进登录组件

再补充一点，在 React 组件的 render() 方法中返回空（null），并不会影响组件的生命周期。

第 7 章

◀ React 列表与 Key ▶

本章介绍如何在 React 框架中创建、转化和使用列表,以及如何通过 Key 来识别元素改变的操作。

7.1 React 列表介绍

在 React 框架中,设计人员可以借助数组实现转化元素列表的渲染操作,同时需要通过 JavaScript 数组(Array)对象的 map()方法来实现。Array 对象的 map()方法返回一个新数组,新数组中的元素为原始数组元素调用自定义函数处理后的值。

下面,我们就先介绍 Array 对象的 map()方法如何使用,请看下面的代码:

【代码 7-1】(详见源代码目录 ch07-react-arr-map.html 文件)

```
01  // TODO: define num
02  const num = [1, 2, 3];
03  console.log(num);
04  // TODO: calculate squrare
05  const square = num.map((num) => num * num);
06  console.log(square);
```

关于【代码 7-1】的说明:

- 第 02 行代码定义了一个常量数组(num),初始化为[1, 2, 3]。
- 第 05 行代码通过 Array 对象的 map()方法返回了一个新常量数组(square),新数组的值为原始数组值的平方,此处的平方运算是通过箭头函数完成的。

测试网页的效果如图 7.1 所示。如图中的箭头所示,浏览器控制台中显示了通过 Array 对象的 map()方法返回的新数组内容。

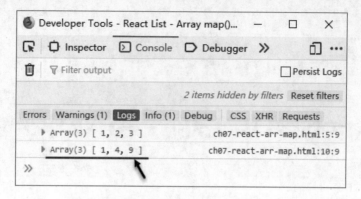

图 7.1 使用 Array.map()方法

在 React 框架中，借助数组转化为元素列表的过程与【代码 7-1】十分类似。下面，我们再看一个通过 Array 对象的 map()方法渲染多个元素组件的代码实例，具体如下：

【代码 7-2】（详见源代码目录 ch07-react-map-list.html 文件）

```
01  <!DOCTYPE html>
02  <html>
03  <head>
04      <meta charset="UTF-8" />
05      <title>React List - Map List</title>
06      <script src="https://unpkg.com/react@16/umd/react.development.js"></script>
07      <script src="https://unpkg.com/react-dom@16/umd/react-dom.development.js"></script>
08      <!-- Don't use this in production: -->
09      <script src="https://unpkg.com/babel-standalone@6.15.0/babel.min.js"></script>
10  </head>
11  <body>
12  <!-- 添加文档主体内容 -->
13  <div id='id-div-react'></div>
14  <script type="text/babel">
15      // TODO: get div
16      var divReact = document.getElementById('id-div-react');
17      // TODO: define const
18      const alpha = ['a', 'e', 'i', 'o', 'u'];
19      // TODO: define li list
20      const alphaList = alpha.map(
21          (alpha) => <li>{alpha}</li>
22      );
23      // TODO: React JSX
24      const reactSpan = (
25          <span>
```

```
26          <h3>React List - Map List</h3>
27          <ul>{alphaList}</ul>
28        </span>
29      );
30      // TODO: React render
31      ReactDOM.render(reactSpan, divReact);
32 </script>
33 </body>
34 </html>
```

关于【代码 7-2】的说明：

- 第 18 行代码定义了一个常量数组（alpha），初始化为元音字母（['a', 'e', 'i', 'o', 'u']）。
- 第 20~22 行代码是关键部分，通过 Array 对象的 map()方法返回了一个列表（<li>标签元素）组件（alphaList）。
- 第 27 行代码将列表组件（alphaList）插入到<ul>标签元素中，组合成一个无序列表并渲染到页面中。

测试网页的效果如图 7.2 所示。如图中的箭头所示，页面中成功显示出了通过 Array 对象的 map()方法返回的列表组件。

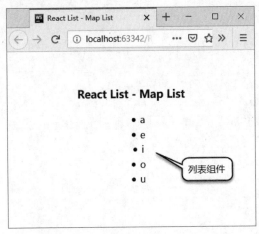

图 7.2　React 列表——渲染多个组件

7.2　基础列表组件

正常情况下，设计人员需要构建一个组件来渲染列表。因此，可以将【代码 7-2】重构成一个组件，通过这个组件接受元音字母的数组作为参数，然后渲染出一个元素列表。关于这个基础列表组件的具体代码如下：

【代码 7-3】（详见源代码目录 ch07-react-list-comp.html 文件）

```html
01  <!DOCTYPE html>
02  <html>
03  <head>
04      <meta charset="UTF-8" />
05      <title>React List - Map List</title>
06      <script src="https://unpkg.com/react@16/umd/react.development.js"></script>
07      <script src="https://unpkg.com/react-dom@16/umd/react-dom.development.js"></script>
08      <!-- Don't use this in production: -->
09      <script src="https://unpkg.com/babel-standalone@6.15.0/babel.min.js"></script>
10  </head>
11  <body>
12  <!-- 添加文档主体内容 -->
13  <div id='id-div-react'></div>
14  <script type="text/babel">
15      // TODO: get div
16      var divReact = document.getElementById('id-div-react');
17      // TODO: function component
18      function MapList(props) {
19          const alpha = props.alpha;
20          const listAlpha = alpha.map(
21              (alpha) => <li>{alpha.toUpperCase()} - {alpha}</li>
22          );
23          return (
24              <ul>
25                  {listAlpha}
26              </ul>
27          );
28      }
29      // TODO: define array const
30      const alpha = ['a', 'e', 'i', 'o', 'u'];
31      // TODO: React JSX
32      const reactSpan = (
33          <span>
34              <h3>React List - Map List</h3>
35              <MapList alpha={alpha} />
36          </span>
37      );
38      // TODO: React render
39      ReactDOM.render(reactSpan, divReact);
```

```
40    </script>
41    </body>
42    </html>
```

关于【代码7-3】的说明：

- 第18~28行代码定义了一个函数组件（MapList），用于实现一个列表组件，具体说明如下：
  - 第20~22行代码是关键部分，通过Array对象的map()方法返回了一个列表（<li>标签元素）组件（listAlpha）。另外，第21行代码通过调用toUpperCase()方法将小写元音字母转换为对应的大写元音字母。
  - 第23~27行代码将列表组件（listAlpha）插入到<ul>标签元素中，组合成一个无序列表。
- 第29行代码定义了一个常量数组（alpha），初始化为元音字母（['a', 'e', 'i', 'o', 'u']）。
- 第35行代码引用函数组件（MapList），并通过属性（alpha）传递给函数组件（MapList）的Props参数值。

测试网页的效果如图7.3所示。如图中的标识所示，页面中成功显示出了通过Array对象的map()方法返回的元音字母（大小写对应）基础列表组件。

图7.3　React列表——基础列表组件

7.3　多级列表组件

本节在前面基础列表组件的基础上，再设计一个多级列表组件，具体的实现代码如下：

【代码 7-4】（详见源代码目录 ch07-react-list-multi-comp.html 文件）

```html
01  <!DOCTYPE html>
02  <html>
03  <head>
04      <meta charset="UTF-8" />
05      <title>React List - Map List</title>
06      <script src="https://unpkg.com/react@16/umd/react.development.js"></script>
07      <script src="https://unpkg.com/react-dom@16/umd/react-dom.development.js"></script>
08      <!-- Don't use this in production: -->
09      <script src="https://unpkg.com/babel-standalone@6.15.0/babel.min.js"></script>
10  </head>
11  <body>
12  <!-- 添加文档主体内容 -->
13  <div id='id-div-react'></div>
14  <script type="text/babel">
15      // TODO: get div
16      var divReact = document.getElementById('id-div-react');
17      // TODO: function component
18      function UserList(props) {
19          const users = props.users;
20          const userName = users.name;
21          const userAge = users.age;
22          const userGender = users.gender;
23          const skills = users.skills;
24          const usersSkills = skills.map(
25              (skills) => <li>{skills}</li>
26          );
27          return (
28              <ul>
29                  <li>Name - {userName}</li>
30                  <li>Age - {userAge}</li>
31                  <li>Gender - {userGender}</li>
32                  <li>Skills -
33                      <ul>
34                          {usersSkills}
35                      </ul>
36                  </li>
37              </ul>
38          );
39      }
40      // TODO: function component
41      function UserComp(props) {
42          const userlist = props.userList;
43          const userlistMap = userlist.map(
44              (userList) => <UserList users={userList}></UserList>
45          );
46          return (
47              <ul>
48                  {userlistMap}
49              </ul>
50          );
```

```
51        }
52        // TODO: define array const
53        const userList = [{
54           name: 'king',
55           age: 18,
56           gender: 'male',
57           skills: ['JavaScript', 'Java', 'Python']
58        },{
59           name: 'tina',
60           age: 12,
61           gender: 'female',
62           skills: ['eat', 'drink', 'sleep']
63        },{
64           name: 'cici',
65           age: 6,
66           gender: 'female',
67           skills: ['play', 'laugh', 'cry']
68        }];
69        // TODO: React JSX
70        const reactSpan = (
71           <span>
72              <h3>React List - Map Multi-List</h3>
73              <UserComp userList={userList} />
74           </span>
75        );
76        // TODO: React render
77        ReactDOM.render(reactSpan, divReact);
78    </script>
79    </body>
80    </html>
```

关于【代码 7-4】的说明：

- 第 53~68 行代码定义了一个 JSON 对象格式的数组（userList），每个 JSON 对象包括了 name、age、gender 和 skills 字段属性，且 skills 还是一个数组格式的字段。这个 JSON 对象的数据格式相对复杂，而我们的目标就是通过解析该对象构建一个 React 多级列表。

- 第 18~39 行代码定义了一个函数组件（UserList），用于实现一个列表组件，具体说明如下：
  ➢ 第 20~23 行代码定义了一组常量，通过 Props 参数获取了 JSON 对象（userList）的 4 个字段属性（name、age、gender 和 skills 字段）。
  ➢ 第 24~26 行代码是比较关键的部分，通过 Array 对象的 map() 方法解析了 skills 字段，并返回了一个列表组件（usersSkills）。
  ➢ 第 27~38 行代码负责将解析得到的 name、age、gender 和 skills 字段组合成一个列表组件（包括一个完整的用户信息）并返回。

- 第 41~51 行代码又定义了一个函数组件（UserComp），并通过引入函数组件（UserList）构建出一个完整的用户列表，具体说明如下：
  ➢ 第 42 行代码通过 Props 参数获取了传递进来的列表对象（userList）。

- ➢ 第 43～45 行代码是关键部分，通过 Array 对象的 map()方法解析了列表对象（userList），并返回了一个完整列表组件（userlistMap）。注意，此时的列表组件（userlistMap）已经包含了第 53～68 行代码所定义的完整的用户信息了。
- ➢ 第 46～50 行代码通过 return()方法返回了经过<ul>标签包裹的列表组件（userlistMap）。

测试网页的效果如图 7.4 所示。页面中成功显示出了通过 Array 对象的 map()方法解析复杂的 JSON 对象格式数据、构建而成的多级用户信息列表组件。

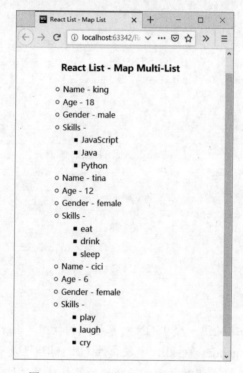

图 7.4　React 列表——多级列表组件

7.4 React Key 介绍

在 React 列表设计中，还有一个非常重要的概念就是"key"，这是 React 框架所特有的知识点。那么，这个"key"具体是个什么概念呢？先不忙解释，我们再回过头去看一下【代码 7-2】的内容。

主要是看一下【代码 7-2】在浏览器中的运行结果，尤其是浏览器控制台的提示信息。下面使用 Firefox 浏览器再次运行测试【代码 7-2】，注意要打开浏览器控制台，具体如图 7.5 所示。如图中的箭头所示，浏览器控制台中显示了一个错误提示（Each child in a list should have a unique "key" prop.），意思是列表中的每一项必须包括一个特殊的"key"属性。

第 7 章 React 列表与 Key

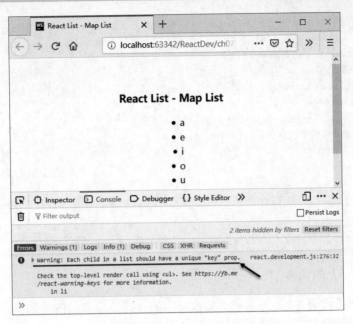

图 7.5 关于 React key 的介绍

既然 React 框架对于列表有这样的规定，我们就尝试为每一个列表项定义一个"key"属性。请看下面的代码实例：

【代码 7-5】（详见源代码目录 ch07-react-list-key.html 文件）

```
01  <!DOCTYPE html>
02  <html>
03  <head>
04      <meta charset="UTF-8" />
05      <title>React List - Map List</title>
06      <script src="https://unpkg.com/react@16/umd/react.development.js"></script>
07      <script src="https://unpkg.com/react-dom@16/umd/react-dom.development.js"></script>
08      <!-- Don't use this in production: -->
09      <script src="https://unpkg.com/babel-standalone@6.15.0/babel.min.js"></script>
10  </head>
11  <body>
12  <!-- 添加文档主体内容 -->
13  <div id='id-div-react'></div>
14  <script type="text/babel">
15      // TODO: define i
16      var i = 0;
17      // TODO: get div
18      var divReact = document.getElementById('id-div-react');
19      // TODO: define const
20      const alpha = ['a', 'e', 'i', 'o', 'u'];
21      // TODO: define li list
22      const alphaList = alpha.map(
```

```
23            (alpha) => <li key={i++}>{alpha}</li>
24        );
25        console.log(alphaList);
26        // TODO: React JSX
27        const reactSpan = (
28            <span>
29                <h3> React List Key - List Key </h3>
30                <ul>{alphaList}</ul>
31            </span>
32        );
33        // TODO: React render
34        ReactDOM.render(reactSpan, divReact);
35    </script>
36    </body>
37 </html>
```

关于【代码 7-5】的说明：

这段代码是在【代码 7-2】的基础上修改而成的，具体增加的内容如下：

- 第 16 行代码定义了一个计数器变量（i），初始化为 0。
- 第 23 行代码在<li>标签元素中增加定义了一个属性（key），属性值为计数器变量（i）的累加值（i++）。

测试网页的效果如图 7.6 所示。如图中的箭头所示，浏览器控制台中的错误信息没有了，说明增加了"key"的定义后通过了 React 框架的检测。

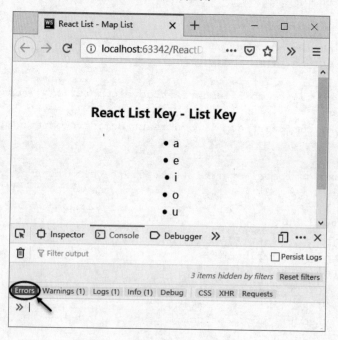

图 7.6　React key 的定义（一）

下面再看一下浏览器控制台中的日志信息窗口，如图 7.7 所示。如图中的箭头和标识所示，浏览器控制台中显示的列表项信息中，包括了"key"属性及其属性值。

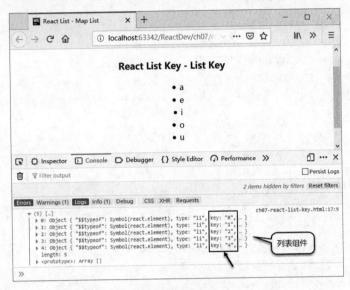

图 7.7　React key 的定义（二）

7.5　React Key 使用

在 React 列表中使用 key，可以帮助 React 框架识别元素的改变（如添加元素或元素被删除，等等）。因此，设计人员应当为每一个元素定义一个确定的标识（key），而且一个元素的 key 最好在其所在的列表中是一个独一无二的字符串。

通常情况下，建议使用 id 值（一般 id 是独一无二的）作为元素的 key，请看下面的代码实例：

【代码 7-6】（详见源代码目录 ch07-react-list-key-id.html 文件）

```
01  <!DOCTYPE html>
02  <html>
03  <head>
04      <meta charset="UTF-8" />
05      <title>React List - Map List</title>
06      <script src="https://unpkg.com/react@16/umd/react.development.js"></script>
07      <script src="https://unpkg.com/react-dom@16/umd/react-dom.development.js"></script>
08      <!-- Don't use this in production: -->
09      <script
```

```
         src="https://unpkg.com/babel-standalone@6.15.0/babel.min.js"></script>
    10  </head>
    11  <body>
    12      <!-- 添加文档主体内容 -->
    13      <div id='id-div-react'></div>
    14      <script type="text/babel">
    15          // TODO: get div
    16          var divReact = document.getElementById('id-div-react');
    17          // TODO: define const
    18          const alpha = ['a', 'e', 'i', 'o', 'u'];
    19          // TODO: define id array
    20          const id = new Array(5);
    21          for(let i=0; i<5; i++) {
    22              id[i] = Math.round((Math.random()*Math.pow(10,10)));
    23          }
    24          // TODO: array map to generate list
    25          const alphaList = alpha.map(
    26              (alpha, index) => <li key={id[index].toString()}>{alpha}</li>
    27          );
    28          console.log(alphaList);
    29          // TODO: React JSX
    30          const reactSpan = (
    31              <span>
    32                  <h3>React List - Map List</h3>
    33                  <ul>{alphaList}</ul>
    34              </span>
    35          );
    36          // TODO: React render
    37          ReactDOM.render(reactSpan, divReact);
    38      </script>
    39  </body>
    40  </html>
```

关于【代码 7-6】的说明：

这段代码是在【代码 7-5】的基础上修改而成的，具体增加的内容如下：

- 第 20 行代码定义了一个数组（id），用于保存一组<li>列表项的 id 标识。
- 第 21～23 行代码通过 for 循环语句初始化了该数组（id），具体就是第 22 行代码通过对生成随机数运算得到的值。
- 第 25～27 行代码通过 Array 对象的 map()方法生成了一组列表（<li>），第 26 行代码在<li>标签元素中增加定义了一个属性（key），属性值为数组（id）中保存的随机数。

测试网页的效果如图 7.8 所示。如图中的箭头所示，"key"属性值为一串随机数，可以用来表示<li>列表项的唯一 id 标识。

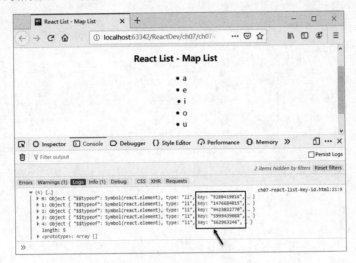

图 7.8　React key 表示 id

7.6　React 通过 Key 提取组件

在 React 列表中，元素的"key"只有放在就近的数组上下文中才会有意义。这句话如何理解呢？我们打个比方，假如在设计 React 列表组件时可以优化提取出一个自定义列表项（<ListItem>）子组件，此时应把"key"保留在这个自定义列表项<ListItem>子组件中，而不是放在最初的<li>标签元素中。

例如，下面代码的写法就是不正确使用"key"的方式，具体如下：

【代码 7-7】（详见源代码目录 ch07-react-list-key-wrong-usage.html 文件）

```
01  <!DOCTYPE html>
02  <html>
03  <head>
04      <meta charset="UTF-8" />
05      <title>React List -  Map List</title>
06      <script src="https://unpkg.com/react@16/umd/react.development.js"></script>
07      <script src="https://unpkg.com/react-dom@16/umd/react-dom.development.js"></script>
08      <!-- Don't use this in production: -->
09      <script src="https://unpkg.com/babel-standalone@6.15.0/babel.min.js"></script>
10  </head>
```

```html
11  <body>
12      <!-- 添加文档主体内容 -->
13      <div id='id-div-react'></div>
14      <script type="text/babel">
15          // TODO: get div
16          var divReact = document.getElementById('id-div-react');
17          // TODO: define array const
18          const alpha = ['a', 'e', 'i', 'o', 'u'];
19          // TODO: define id array
20          const id = new Array(5);
21          for(let i=0; i<5; i++) {
22              id[i] = Math.round((Math.random()*Math.pow(10,10)));
23          }
24          // TODO: function component - list item
25          function ListItem(props) {
26              const alpha = props.alpha;
27              const index = props.index;
28              return (
29                  // TODO：错误写法，不需要在这里指定 key
30                  <li key={id[index].toString()}>{alpha}</li>
31              );
32          }
33          // TODO: function component - list comp
34          function AlphaList(props) {
35              const alpha = props.alpha;
36              const listItems = alpha.map((alpha, index) =>
37                  // TODO: 错误写法，元素的 key 应该在这里指定
38                  <ListItem alpha={alpha} index={index} />
39              );
40              return (
41                  <ul>
42                      {listItems}
43                  </ul>
44              );
45          }
46          // TODO: React JSX
47          const reactSpan = (
48              <span>
49                  <h3>React List - Map List</h3>
50                  <AlphaList alpha={alpha} />
51              </span>
52          );
53          // TODO: React render
```

```
54        ReactDOM.render(reactSpan, divReact);
55    </script>
56  </body>
57 </html>
```

关于【代码 7-7】的说明：

- 第 25~32 行代码定义了一个函数组件（ListItem），具体说明如下：
  - 该函数组件相当于是子组件，用于定义列表项。
  - 在第 30 行代码中，在<li>标签元素中增加定义了一个属性（key），相当于元素的 id 值。
  - 另外，第 30 行代码定义 "key" 的做法是不正确的，读者会在后面的运行测试结果中看到。
- 第 34~45 行代码定义了另一个函数组件（AlphaList），第 38 行代码通过引用函数组件（<ListItem>，相当于列表项）组成一个列表组件。

测试网页的效果如图 7.9 所示。虽然页面中成功显示出了元音字母的列表，但从浏览器控制台中的错误提示信息可以看到缺少列表 "key" 的定义。可是，在第 30 行代码中明明有定义 "key"，为什么会有错误提示信息呢？

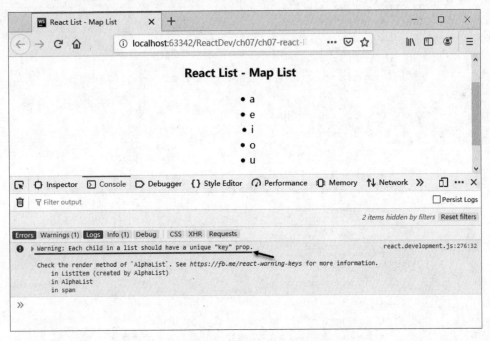

图 7.9　React key 错误的用法

答案是代码中 "key" 的定义位置发生错误了，下面是正确使用 "key" 方式的代码写法：

【代码 7-8】（详见源代码目录 ch07-react-list-key-right-usage.html 文件）

```
01  <!DOCTYPE html>
```

```html
02  <html>
03  <head>
04      <meta charset="UTF-8" />
05      <title>React List - Map List</title>
06      <script src="https://unpkg.com/react@16/umd/react.development.js"></script>
07      <script src="https://unpkg.com/react-dom@16/umd/react-dom.development.js"></script>
08      <!-- Don't use this in production: -->
09      <script src="https://unpkg.com/babel-standalone@6.15.0/babel.min.js"></script>
10  </head>
11  <body>
12  <!-- 添加文档主体内容 -->
13  <div id='id-div-react'></div>
14  <script type="text/babel">
15      // TODO: get div
16      var divReact = document.getElementById('id-div-react');
17      // TODO: define array const
18      const alpha = ['a', 'e', 'i', 'o', 'u'];
19      // TODO: define id array
20      const id = new Array(5);
21      for(let i=0; i<5; i++) {
22          id[i] = Math.round((Math.random()*Math.pow(10,10)));
23      }
24      // TODO: function component
25      function ListItem(props) {
26          const alpha = props.alpha;
27          return (
28              // TODO: 正确写法，这里不需要指定 key
29              <li>{alpha}</li>
30          );
31      }
32      // TODO: function component
33      function AlphaList(props) {
34          const alpha = props.alpha;
35          const listItems = alpha.map((alpha, index) =>
36              // TODO: 正确写法，key 应该在数组的上下文中被指定
37              <ListItem key={id[index].toString()} alpha={alpha} />
38          );
39          console.log(listItems);
40          return (
41              <ul>
```

```
42                {listItems}
43            </ul>
44        );
45    }
46    // TODO: React JSX
47    const reactSpan = (
48        <span>
49            <h3>React List - Map List</h3>
50            <AlphaList alpha={alpha} />
51        </span>
52    );
53    // TODO: React render
54    ReactDOM.render(reactSpan, divReact);
55 </script>
56 </body>
57 </html>
```

关于【代码 7-8】的说明：

这段代码主要就是将【代码 7-7】中定义 "key" 的方式改变了，具体说明如下：

- 在【代码 7-7】的第 30 行代码中，在<li>标签元素中定义了属性（key）；而在【代码 7-8】的第 29 行代码中，则删除了<li>标签元素中定义的属性（key）。
- 然后，【代码 7-8】将属性（key）的定义改写到了第 37 行代码中。

测试网页的效果如图 7.10 所示。如图中的标识所示，浏览器控制台中的错误信息没有了，而且列表项的属性 "key" 值也正确显示出来了。

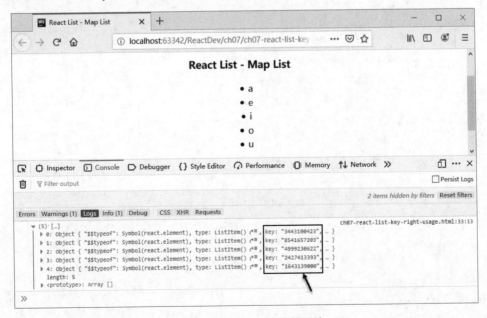

图 7.10　React key 正确的用法

7.7 React Key 局部唯一性

在 React 列表中，元素列表项中定义的 "key" 值在其兄弟节点之间必须是独一无二的。但是，"key" 值并不需要全局也是唯一的，这句话该怎么理解呢？简单描述一下，就是对于同级的两个或多个列表组件中元素列表项所定义的 "key"，就可以定义为相同的 "key" 值。下面，我们看一个具体的代码实例：

【代码 7-9】（详见源代码目录 ch07-react-list-key-unique.html 文件）

```
01  <!DOCTYPE html>
02  <html>
03  <head>
04      <meta charset="UTF-8" />
05      <title>React List - Key Unique</title>
06      <script src="https://unpkg.com/react@16/umd/react.development.js"></script>
07      <script src="https://unpkg.com/react-dom@16/umd/react-dom.development.js"></script>
08      <!-- Don't use this in production: -->
09      <script src="https://unpkg.com/babel-standalone@6.15.0/babel.min.js"></script>
10  </head>
11  <body>
12  <!-- 添加文档主体内容 -->
13  <div id='id-div-react'></div>
14  <script type="text/babel">
15      // TODO: get div
16      var divReact = document.getElementById('id-div-react');
17      // TODO: define array const
18      const arrAlpha = [
19          {
20              id: 1,
21              alpha: 'a'
22          },
23          {
24              id: 2,
25              alpha: 'b'
26          },
27          {
28              id: 3,
29              alpha: 'c'
30          }
```

```
31          ];
32          // TODO: function component
33          function AlphaList(props) {
34              const arrAlpha = props.alpha;
35              const listAlpha = arrAlpha.map(
36                  (arrAlpha) => <li key={arrAlpha.id}>{arrAlpha.alpha}</li>
37              );
38              console.log(listAlpha);
39              const listAlphaUpper = arrAlpha.map(
40                  (arrAlpha) => <li key={arrAlpha.id}>{arrAlpha.alpha.toUpperCase()}</li>
41              );
42              console.log(listAlphaUpper);
43              return (
44                  <span>
45                      <ul>
46                          {listAlpha}
47                      </ul>
48                      <ul>
49                          {listAlphaUpper}
50                      </ul>
51                  </span>
52              );
53          }
54          // TODO: React JSX
55          const reactSpan = (
56              <span>
57                  <h3>React List - Key Unique</h3>
58                  <AlphaList alpha={arrAlpha}/>
59              </span>
60          );
61          // TODO: React render
62          ReactDOM.render(reactSpan, divReact);
63      </script>
64  </body>
65  </html>
```

关于【代码7-9】的说明：

- 第18~31行代码定义了一个JSON对象格式的数组（arrAlpha），每个JSON对象包括了"id"和"alpha"字段属性。我们的目标就是通过解析该JSON数组对象，同时构建两个具有相同"key"值的列表。

- 第33~53行代码定义了一个函数组件（AlphaList），具体说明如下：

➢ 第 35~37 行代码中，通过 Array 对象的 map()方法返回了第一个列表（<li>标签元素）组件（listAlpha），并通过 "id" 属性定义了 "key" 值。
➢ 第 39~41 行代码中，通过 Array 对象的 map()方法返回了第二个列表（<li>标签元素）组件（listAlphaUpper），同样通过 "id" 属性定义了 "key" 值；另外，通过 toUpperCase()方法将列表项值的小写元音字母转换为了大写元音字母。
➢ 第 43~52 行代码同时将这两个列表组件（listAlpha）和（listAlphaUpper）渲染到页面中显示。

测试网页的效果如图 7.11 所示。如图中的标识所示，浏览器控制台中给出了小写元音字母列表与大写元音字母列表的 "key" 值，二者是完全相同的。

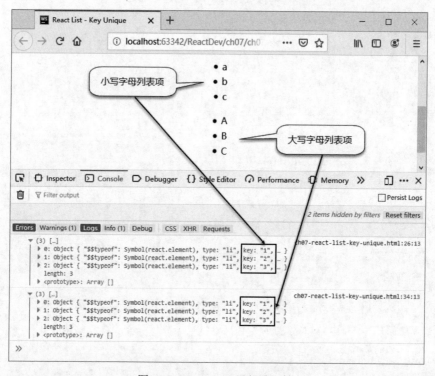

图 7.11　React Key 局部唯一性

7.8　React Key 有效性

React Key 还有一个很特殊的读取有效性，就是 "key" 值仅仅会传递信息给 React 框架，但用户的自定义组件是无法读取的。如果需要在组件中使用属性 "key" 的值，那么只能通过其他属性名显式传递 "key" 值来变通了。

下面，我们通过一个具体的代码实例介绍 React Key 的读取有效性，具体如下：

【代码 7-10】（详见源代码目录 ch07-react-list-key-read.html 文件）

```
01  <!DOCTYPE html>
02  <html>
03  <head>
04      <meta charset="UTF-8" />
05      <title>React List - Key Unique</title>
06      <script src="https://unpkg.com/react@16/umd/react.development.js"></script>
07      <script src="https://unpkg.com/react-dom@16/umd/react-dom.development.js"></script>
08      <!-- Don't use this in production: -->
09      <script src="https://unpkg.com/babel-standalone@6.15.0/babel.min.js"></script>
10  </head>
11  <body>
12  <!-- 添加文档主体内容 -->
13  <div id='id-div-react'></div>
14  <script type="text/babel">
15      // TODO: get div
16      var divReact = document.getElementById('id-div-react');
17      // TODO: define array const
18      const arrAlpha = [
19          {
20              id: 1,
21              alpha: 'a'
22          },
23          {
24              id: 2,
25              alpha: 'b'
26          },
27          {
28              id: 3,
29              alpha: 'c'
30          }
31      ];
32      // TODO: function component
33      function AlphaList(props) {
34          const arrAlpha = props.alpha;
35          const listAlpha = arrAlpha.map(
36              (arrAlpha) => <li key={arrAlpha.id}>{arrAlpha.alpha}</li>
37          );
38          console.log(listAlpha);
39          const listAlphaUpper = arrAlpha.map(
40              (arrAlpha) => <li key={arrAlpha.id}>{arrAlpha.alpha.toUpperCase()}</li>
41          );
42          console.log(listAlphaUpper);
43          return (
44              <span>
45                  <ul>
```

```
46                    {listAlpha}
47                </ul>
48                <ul>
49                    {listAlphaUpper}
50                </ul>
51            </span>
52        );
53    }
54    // TODO: React JSX
55    const reactSpan = (
56        <span>
57            <h3>React List - Key Unique</h3>
58            <AlphaList alpha={arrAlpha}/>
59        </span>
60    );
61    // TODO: React render
62    ReactDOM.render(reactSpan, divReact);
63 </script>
64 </body>
65 </html>
```

关于【代码 7-10】的说明：

- 第 18～31 行代码定义了一个 JSON 对象格式的数组（arrAlpha），每个 JSON 对象包括了"id"和"alpha"字段属性。我们的目标就是通过解析该 JSON 数组对象，同时构建两个具有相同"key"值的列表。
- 第 33～53 行代码定义了一个函数组件（AlphaList），具体说明如下：
  - 第 35～37 行代码中，通过 Array 对象的 map() 方法返回了第一个列表（<li>标签元素）组件（listAlpha），并通过"id"属性定义了"key"值。
  - 第 39～41 行代码中，通过 Array 对象的 map() 方法返回了第二个列表（<li>标签元素）组件（listAlphaUpper），同样通过"id"属性定义了"key"值；另外，通过 toUpperCase() 方法将列表项值的小写元音字母转换为了大写元音字母。
  - 第 43～52 行代码同时将这两个列表组件（listAlpha）和（listAlphaUpper）渲染到页面中显示。

测试网页的效果如图 7.12 所示。如图中的标识所示，浏览器控制台中给出了小写元音字母列表与大写元音字母列表的"key"值，二者是完全相同的。

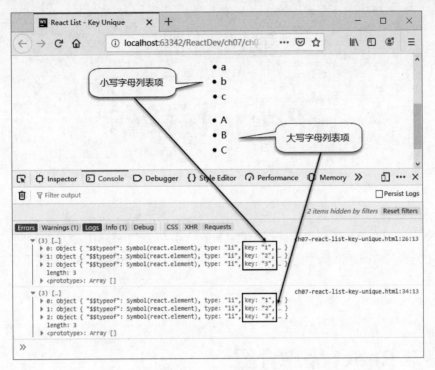

图 7.12　React key 局部唯一性

第 8 章

◀ React 表单 ▶

本章介绍如何在 React 框架中使用表单（Form），以及与 HTML 表单的异同。React 表单中的 DOM 元素与 React 框架中的其他 DOM 元素有所不同，因为表单元素需要保留一些内部状态（State）。

8.1 React 表单介绍

React 框架中的表单（Form）具有与 HTML 表单相同的默认行为，也就是说如果你在 React 框架中执行 HTML 表单，代码仍然是有效的。为了验证上面的描述，我们通过一段简单的 HTML 表单代码验证一下：

【代码 8-1】（详见源代码目录 ch08-react-html-form.html 文件）

```
01  <!DOCTYPE html>
02  <html>
03  <head>
04      <meta charset="UTF-8" />
05      <title>React Form - HTML Form</title>
06      <script src="https://unpkg.com/react@16/umd/react.development.js"></script>
07      <script src="https://unpkg.com/react-dom@16/umd/react-dom.development.js"></script>
08      <!-- Don't use this in production: -->
09      <script src="https://unpkg.com/babel-standalone@6.15.0/babel.min.js"></script>
10  </head>
11  <body>
12      <!-- 添加文档主体内容 -->
13      <h3>React Form - HTML Form</h3>
14      <div id='id-div-react'>
15          <form name="frmName">
```

```
16          <label>Name:</label><br/>
17          <input type="text" name="username" value="" /><br/><br/>
18          <input type="submit" value="Submit" />
19      </form>
20  </div>
21  <script type="text/babel">
22      // TODO: init Form fields
23      document.forms[0].username.value = "king";
24  </script>
25  </body>
26  </html>
```

关于【代码 8-1】的说明：

- 整段代码是在 React 框架下完成的，脚本标签（<script type="text/babel">）的类型定义的也是"babel"。
- 第 15~19 行代码定义了一个表单（Form），内含一个"用户名"文本域。
- 第 23 行代码通过 Document 对象对表单中的"用户名"文本域进行了初始化操作，这里是通过 JS 方式实现的。

测试网页的效果如图 8.1 所示。如图中的箭头所示，表单中的"用户名"文本域中显示了第 23 行代码初始化的文本（"king"），说明在 React 框架下执行 HTML 表单代码是有效的。

图 8.1 React 框架中使用 HTML 表单

下面，再看一下如何将【代码 8-1】通过 React 的方式来实现，具体的代码如下：

【代码 8-2】（详见源代码目录 ch08-react-form.html 文件）

```
01  <!DOCTYPE html>
02  <html>
03  <head>
04      <meta charset="UTF-8" />
05      <title>React Form - React Form</title>
```

```
06        <script src="https://unpkg.com/react@16/umd/react.development.js"></script>
07        <script src="https://unpkg.com/react-dom@16/umd/react-dom.development.js"></script>
08        <!-- Don't use this in production: -->
09        <script src="https://unpkg.com/babel-standalone@6.15.0/babel.min.js"></script>
10    </head>
11    <body>
12        <!-- 添加文档主体内容 -->
13        <div id='id-div-react'></div>
14        <script type="text/babel">
15            // TODO: get div
16            var divReact = document.getElementById('id-div-react');
17            // TODO: function component
18            function FormComp(props) {
19                return (
20                    <form>
21                        <label>Name:</label><br/>
22                        <input type="text" name="name" value={props.username} /><br/><br/>
23                        <input type="submit" value="Submit" />
24                    </form>
25                );
26            }
27            // TODO: React JSX
28            const reactSpan = (
29                <span>
30                    <h3>React Form - React Form</h3>
31                    <FormComp username="king" />
32                </span>
33            );
34            // TODO: React render
35            ReactDOM.render(reactSpan, divReact);
36        </script>
37    </body>
38 </html>
```

关于【代码 8-2】的说明：

- 第 18~26 行代码定义了一个表单函数组件（FormComp），其中第 20~24 行代码是表单的定义。另外，这个表单函数组件（FormComp）通过 Props 参数传递 "用户名" 信息，在第 22 行代码中 "用户名" 信息显示在文本输入框中。

- 第 31 行代码将表单函数组件（FormComp）渲染到页面中，通过 "username" 属性初始化用户名（"king"）。

测试网页的效果如图 8.2 所示。如图中的箭头所示，表单中的"用户名"文本域中显示了第 21 行代码初始化的属性值（"king"），效果与图 8.1 所示的是完全一样的。

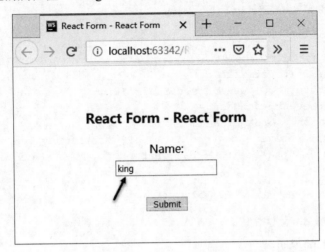

图 8.2　React 表单

8.2　表单受控组件

传统 HTML 表单中的元素（如文本输入框<input>、下拉菜单<select>和按钮<button>，等等）通常是自己维护自己的"状态"，根据用户输入的改变进行"状态"更新。下面，我们通过一个简单的代码实例验证一下上面的描述，具体如下：

【代码 8-3】（详见源代码目录 ch08-react-html-form-control.html 文件）

```
01  <!DOCTYPE html>
02  <html>
03  <head>
04      <meta charset="UTF-8" />
05      <title>React Form - HTML Form</title>
06      <script src="https://unpkg.com/react@16/umd/react.development.js"></script>
07      <script src="https://unpkg.com/react-dom@16/umd/react-dom.development.js"></script>
08      <!-- Don't use this in production: -->
09      <script src="https://unpkg.com/babel-standalone@6.15.0/babel.min.js"></script>
10  </head>
11  <body>
12      <!-- 添加文档主体内容 -->
```

```
13    <h3>React Form - HTML Form Control</h3>
14    <div id='id-div-react'>
15        <form name="frmName">
16            <label>Name:</label><br/>
17            <input type="text" name="username" value="" onChange="on_username_change();"/>
18            <input type="submit" value="Submit" />
19        </form>
20    </div>
21    <script type="text/babel">
22        // TODO: init Form fields
23        document.forms[0].username.value = "king";
24        // TODO: event - onChange
25        function on_username_change() {
26            var vUsername = document.forms[0].username.value;
27            console.log("Username has changed to '" + vUsername + "'.");
28        }
29    </script>
30    </body>
31 </html>
```

关于【代码8-3】的说明：

- 第15~19行代码定义了一个表单（Form），内含一个"用户名"文本域。该文本域可以接受用户输入的用户名，那么该如何维护该文本域呢？传统的方式就是为该文本域定义事件监控方法，比如第17行代码定义的"onChange"事件，可以监控输入框内容的变化。
- 第25~28行代码是"onChange"事件处理方法（on_username_change()）的实现过程，该方法会将文本框内容的变化输出到浏览器控制台中。

测试网页的效果如图8.3所示。如图中的箭头所示，传统HTML表单在React框架下可以使用传统的JavaScript代码方式进行正常维护。

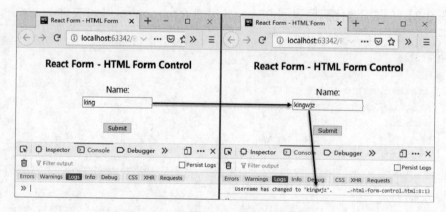

图8.3　HTML表单监控

但是，React框架下表单元素的可变状态（mutable state）通常保存在组件的state状态中，由

React 框架来控制 state 状态的属性值。还是通过一个简单的代码实例来验证一下上面的描述，具体如下：

【代码 8-4】（详见源代码目录 ch08-react-form-control-state.html 文件）

```
01  <!DOCTYPE html>
02  <html>
03  <head>
04      <meta charset="UTF-8" />
05      <title>React Form - React Form</title>
06      <script src="https://unpkg.com/react@16/umd/react.development.js"></script>
07      <script src="https://unpkg.com/react-dom@16/umd/react-dom.development.js"></script>
08      <!-- Don't use this in production: -->
09      <script src="https://unpkg.com/babel-standalone@6.15.0/babel.min.js"></script>
10  </head>
11  <body>
12  <!-- 添加文档主体内容 -->
13  <div id='id-div-react'></div>
14  <script type="text/babel">
15      // TODO: get div
16      var divReact = document.getElementById('id-div-react');
17      // TODO: define ES6 Class Component
18      class FormComp extends React.Component {
19          constructor(props) {
20              super(props);
21              this.state = {username: 'king'};
22          }
23          // TODO: render form
24          render() {
25              return (
26                  <form>
27                      <label>Name:</label><br/>
28                      <input type="text" name="name" value={this.state.username} />
29                      <input type="submit" value="Submit" />
30                  </form>
31              );
32          }
33      }
34      // TODO: React JSX
35      const reactSpan = (
```

```
36        <span>
37            <h3>React Form - Form Control State</h3>
38            <FormComp />
39        </span>
40    );
41    // TODO: React render
42    ReactDOM.render(reactSpan, divReact);
43 </script>
44 </body>
45 </html>
```

关于【代码 8-4】的说明：

- 这段代码主要是通过 ES6 Class 组件的方式构建了一个 React 表单，然后通过 state 状态属性来绑定表单域中的文本输入框。
- 第 18～33 行代码定义了一个表单类组件（FormComp），具体内容如下：
  > 第 19～22 行代码的构造方法中，第 21 行代码定义了一个 state 状态属性（username），并初始化了属性值（'king'）。
  > 第 24～32 行代码通过 render()方法渲染了一个表单，其中第 28 行代码定义的文本域中，将 state 状态属性（username）绑定到文本域的"value"属性上。

测试网页的效果如图 8.4 所示。如图中的箭头所示，当用户尝试在文本输入框中改变内容时，会发现是无法进行有效操作的。原因就是在第 28 行代码中，React 框架将 state 状态属性绑定了该文本域。我们可以回想下前面关于 React state 的介绍，state 状态属性是无法人为进行修改的，而只能使用 setState()方法进行修改。

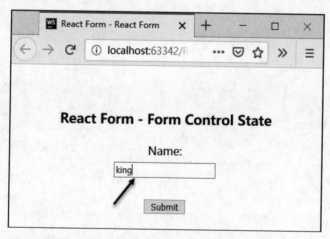

图 8.4　React State 绑定表单域

因此，React 表单的可变状态（mutable state）是需要保存在组件的 state 状态属性中的，并且是只能通过使用 setState()方法进行维护更新的。不过我们正好可以因势利导，通过使用 state 状态属性成为 React 表单域元素的"唯一数据源"，进而控制 React 表单组件"状态"的更新与维护。React 框架为以该方式控制的表单域元素起了一个高端的称谓——React 表单"受控组件"。

下面将【代码 8-4】做进一步完善，以"受控组件"方式实现一个 React 表单，具体代码如下：

【代码 8-5】（详见源代码目录 ch08-react-form-control-comp.html 文件）

```
01  <!DOCTYPE html>
02  <html>
03  <head>
04      <meta charset="UTF-8" />
05      <title>React Form - React Form</title>
06      <script src="https://unpkg.com/react@16/umd/react.development.js"></script>
07      <script src="https://unpkg.com/react-dom@16/umd/react-dom.development.js"></script>
08      <!-- Don't use this in production: -->
09      <script src="https://unpkg.com/babel-standalone@6.15.0/babel.min.js"></script>
10  </head>
11  <body>
12  <!-- 添加文档主体内容 -->
13  <div id='id-div-react'></div>
14  <script type="text/babel">
15      // TODO: get div
16      var divReact = document.getElementById('id-div-react');
17      // TODO: define ES6 Class Component
18      class FormComp extends React.Component {
19          // TODO: Component Constructor
20          constructor(props) {
21              super(props);
22              this.state = {username: 'king'};
23              this.handleChange = this.handleChange.bind(this);
24          }
25          // TODO: handle event
26          handleChange(event) {
27              let eUsername = event.target.value;
28              this.setState({username: eUsername});
29              console.log("Username has changed to '" + eUsername + "'.");
30          }
31          // TODO: render form
32          render() {
33              return (
34                  <form>
35                      <label>Name:</label><br/>
36                      <input type="text" name="name" value={this.state.username}
```

```
onChange={this.handleChange}/>
37                      <input type="submit" value="Submit" />
38              </form>
39          );
40      }
41  }
42  // TODO: React JSX
43  const reactSpan = (
44      <span>
45          <h3>React Form - Form Control Comp</h3>
46          <FormComp />
47      </span>
48  );
49  // TODO: React render
50  ReactDOM.render(reactSpan, divReact);
51 </script>
52 </html>
```

关于【代码 8-5】的说明：

- 这段代码主要是在【代码 8-4】的基础上进行了完善，为文本输入框添加了"onChange"事件处理方法。
- 第 18～41 行代码定义了一个表单类组件（FormComp），具体内容如下：
  - 第 22 行代码定义了一个 state 状态属性（username），并绑定第 36 行代码定义的文本输入框中的"value"属性。
  - 第 36 行代码定义的文本输入框中，添加了"onChange"事件处理方法（handleChange）；根据 React 框架的事件规范，该事件处理方法（handleChange）需要在构造方法中通过 bind() 方法进行了绑定（见第 23 行代码）。
  - 第 26～30 行代码是事件处理方法（handleChange）实现过程，关键之处就是使用 setState() 方法来更新 state 状态属性，并同步更新到文本输入框的内容。

测试网页，页面初始效果如图 8.5 所示。

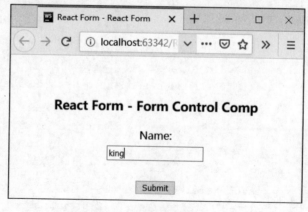

图 8.5　React 表单受控组件（一）

我们可以尝试在文本输入框中改变内容，然后打开浏览器控制台查看一下效果变化，具体如图 8.6 所示。如图中的箭头所示，当用户尝试在文本输入框中改变内容时，文本输入框的内容是完全可以更改的（区别于【代码 8-4】）。同时，浏览器控制台中可以监控到用户的输入日志。

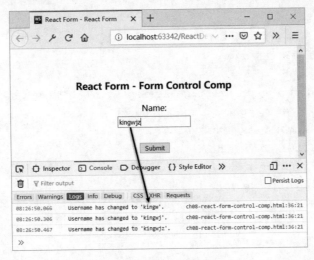

图 8.6　React 表单受控组件（二）

8.3　强制转换大写字母

在前面章节中，我们详细介绍如何实现 React 框架的文本输入框"受控组件"。那么，文本输入框"受控组件"有什么特别之处呢？其实，通过文本输入框"受控组件"是可以实现既有趣又实用的页面功能的。

在很多国外门户网站的用户注册模块中，姓氏（Last Name）和名字（First Name）是只接受大写字母输入的（算是一种文化约束）。在传统方式下，是可以通过 JavaScript 脚本代码实现该功能的。下面，我们看一下通过 React 文本输入框"受控组件"是如何实现这个功能的：

【代码 8-6】（详见源代码目录 ch08-react-form-name-upperCase.html 文件）

```
01  <!DOCTYPE html>
02  <html>
03  <head>
04      <meta charset="UTF-8" />
05      <title>React Form - React Form</title>
06      <script src="https://unpkg.com/react@16/umd/react.development.js"></script>
07      <script src="https://unpkg.com/react-dom@16/umd/react-dom.development.js"></script>
08      <!-- Don't use this in production: -->
```

```
09      <script src="https://unpkg.com/babel-standalone@6.15.0/babel.min.js"></script>
10    </head>
11    <body>
12      <!-- 添加文档主体内容 -->
13      <div id='id-div-react'></div>
14      <script type="text/babel">
15        // TODO: get div
16        var divReact = document.getElementById('id-div-react');
17        // TODO: define ES6 Class Component
18        class FormComp extends React.Component {
19          // TODO: Component Constructor
20          constructor(props) {
21            super(props);
22            this.state = {
23              firstname: '',
24              lastname: ''
25            };
26            this.handleFirstNameChange = this.handleFirstNameChange.bind(this);
27            this.handleLastNameChange = this.handleLastNameChange.bind(this);
28          }
29          // TODO: handle event
30          handleFirstNameChange(event) {
31            let targetValue = event.target.value;
32            let targetValueUpper = targetValue.toUpperCase();
33            this.setState({firstname: targetValueUpper});
34            console.log("'" + targetValue + "' has changed to '" + targetValueUpper + "'.");
35          }
36          handleLastNameChange(event) {
37            let targetValue = event.target.value;
38            let targetValueUpper = targetValue.toUpperCase();
39            this.setState({lastname: targetValueUpper});
40            console.log("'" + targetValue + "' has changed to '" + targetValueUpper + "'.");
41          }
42          // TODO: render form
43          render() {
44            return (
45              <form>
46                <label>First Name(Only Uppercase):</label>
```

```
47                <input type="text" name="firstname"
48                       value={this.state.firstname}
49                       onChange={this.handleFirstNameChange} />
50                <label>Last Name(Only Uppercase):</label><br/>
51                <input type="text" name="firstname"
52                       value={this.state.lastname}
53                       onChange={this.handleLastNameChange} />
54                <input type="submit" value="Submit" />
55            </form>
56        );
57    }
58 }
59 // TODO: React JSX
60 const reactSpan = (
61     <span>
62         <h3>React Form - Accept UpperCase</h3>
63         <FormComp />
64     </span>
65 );
66 // TODO: React render
67 ReactDOM.render(reactSpan, divReact);
68 </script>
69 </body>
70 </html>
```

关于【代码 8-6】的说明：

- 第 18~58 行代码定义了一个表单类组件（FormComp），具体内容如下：
  - 先看第 43~57 行代码，通过 render() 方法渲染了一个表单；其中，第 47~49 行代码和第 51~53 行代码通过 <input> 标签元素定义了两个文本输入框"受控组件"，分别用于用户输入 "firstname" 和 "lastname"；两个文本输入框均定义了 "value" 属性（绑定了 state 状态属性，见下文分析）和 "onChange" 事件处理方法（handleFirstNameChange 和 handleLastNameChange）。
  - 在第 20~28 行代码定义的构造方法中，定义了两个 state 状态属性（firstname 和 lastname）分别绑定给第 48 行和第 52 行代码定义的两个文本输入框中的 "value" 属性；还有就是将两个 "onChange" 事件处理方法（handleFirstNameChange 和 handleLastNameChange）通过 bind() 方法进行了绑定。
  - 第 30~35 行代码和第 36~41 行代码是 "onChange" 事件处理方法（handleFirstNameChange 和 handleLastNameChange）的实现过程，先通过调用 JavaScript 的 toUpperCase() 方法将用户输入的姓名强制转换为大写字母，然后通过 setState() 方法将转换的大写姓名用于更新 state 状态属性（firstname 和 lastname），并同步更新到文本输入框的内容。

测试网页的效果如图 8.7 和图 8.8 所示。如图中的箭头所示，从浏览器控制台中的输入日志可以看到，用户输入的小写字母姓名被强制转换为大写字母格式的姓名了。

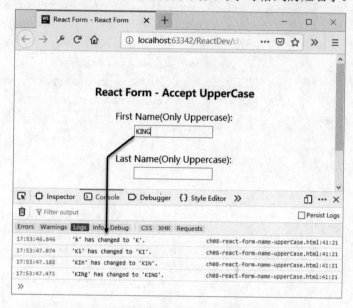

图 8.7　强制转换大写字母（一）

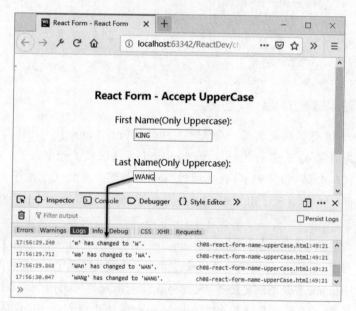

图 8.8　强制转换大写字母（二）

8.4 校验手机号码格式

借助 React 文本输入框"受控组件",还可以很轻松地实现在文本输入框中校验文本输入的格式(如长度、格式,等等)。举例来说,这个功能对于手机号码格式输入的应用场景就很实用。

下面,再看通过 React 文本输入框"受控组件"实现校验手机号码输入长度的代码实例,具体如下:

【代码 8-7】(详见源代码目录 ch08-react-form-tel-length.html 文件)

```
01   <!DOCTYPE html>
02   <html>
03   <head>
04       <meta charset="UTF-8" />
05       <title>React Form - React Form</title>
06       <script src="https://unpkg.com/react@16/umd/react.development.js"></script>
07       <script src="https://unpkg.com/react-dom@16/umd/react-dom.development.js"></script>
08       <!-- Don't use this in production: -->
09       <script src="https://unpkg.com/babel-standalone@6.15.0/babel.min.js"></script>
10   </head>
11   <body>
12   <!-- 添加文档主体内容 -->
13   <div id='id-div-react'></div>
14   <script type="text/babel">
15       // TODO: get div
16       var divReact = document.getElementById('id-div-react');
17       // TODO: define ES6 Class Component
18       class FormComp extends React.Component {
19           // TODO: Component Constructor
20           constructor(props) {
21               super(props);
22               this.state = {
23                   tel: '',
24                   info: '138****8000'
25               };
26               this.handleTelChange = this.handleTelChange.bind(this);
27           }
28           // TODO: handle event
29           handleTelChange(event) {
30               let targetTel = event.target.value;
31               let finalTel = targetTel;
32               if(targetTel.length < 11) {
33                   this.setState({info: 'pls go on...'});
34               } else if(targetTel.length == 11) {
```

```
35                this.setState({info: 'right number.'});
36            } else {
37                finalTel = targetTel.substr(0, 11);
38            }
39            this.setState({tel: finalTel});
40        }
41        // TODO: render form
42        render() {
43            return (
44                <form>
45                    <label>Mobile Phone(+86):</label><br/>
46                    <input type="text" name="tel"
47                        value={this.state.tel}
48                        onChange={this.handleTelChange} />
49                    <input type="text" name="info" className="css-input-info"
50                        value={this.state.info} readOnly/><br/><br/>
51                    <input type="submit" value="Submit" />
52                </form>
53            );
54        }
55    }
56    // TODO: React JSX
57    const reactSpan = (
58        <span>
59            <h3>React Form - Mobile Phone Length</h3>
60            <FormComp />
61        </span>
62    );
63    // TODO: React render
64    ReactDOM.render(reactSpan, divReact);
65 </script>
66 </body>
67 </html>
```

关于【代码 8-7】的说明：

- 先看第 42~54 行代码，通过 render() 方法渲染了一个表单，具体内容如下：
  - 第 46~48 行代码通过<input>标签元素定义了第一个文本输入框"受控组件"，用于用户输入手机号码；其中，"value"属性绑定了 state 状态属性（tel），并添加了"onChange"事件处理方法（handleTelChange）。
  - 第 49~51 行代码通过<input>标签元素定义了第二个文本输入框"受控组件"，用于输出提示信息；其中，"value"属性绑定了 state 状态属性（info），另外还指定了只读（readOnly）属性。
- 第 20~27 行代码定义的构造方法中，定义了两个 state 状态属性（tel 和 info），分别绑定给第 47 行和第 50 行代码定义的两个文本输入框中的"value"属性；还可以将"onChange"事件处理方法（handleTelChange）通过 bind() 方法进行了绑定。
- 这段代码的核心部分是第 29~40 行代码实现的"onChange"事件处理方法

（handleTelChange），具体就是对用户输入手机号码的长度进行了判断和处理。
- 第 30 行代码先通过"event"对象获取了用户输入的内容。
- 第 32～38 行代码通过 if...else...条件语句检测用户输入手机号码的长度（国内的手机号码长度为 11 位），如果长度小于 11 位会提示用户继续输入，如果等于 11 位则提示号码正确，而如果用户继续输入则会强制禁止用户输入（见第 37 行代码的 JavaScript 脚本处理方式）。
- 第 33 行和第 35 行代码通过 setState()方法将提示信息更新到 state 状态属性（tel），并同步更新到第 49～51 行代码定义的只读文本输入框。
- 第 39 行代码使用 setState()方法将用户输入的手机号码用于更新 state 状态属性（tel），并同步更新到第 46～48 行代码定义的文本输入框的内容。

测试网页，页面初始效果如图 8.9 所示。页面中的提示信息位置显示了正确的手机号码长度样式。

图 8.9　校验手机号码长度（一）

下面，我们可以尝试在文本输入框中输入手机号码，页面效果如图 8.10 所示。当用户在文本输入框中输入手机号码的过程中，提示信息位置显示了"pls go on..."表明号码长度还要继续输入。

图 8.10　校验手机号码长度（二）

当我们在文本输入框中输入到正确的手机号码长度时，页面效果如图 8.11 所示。当用户在文本输入框中输入到 11 位的手机号码时，提示信息位置显示了"right number."表明号码长度正确。此时，如果用户继续输入，就会被强制禁止无法执行的。

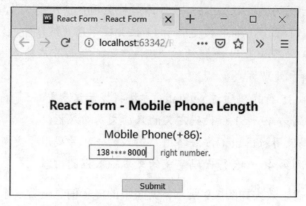

图 8.11　校验手机号码长度（三）

上面是对手机号码的长度进行了校验，其实可以进一步对手机号码的格式进行校验，这就需要用到正则表达式的功能了。国内的手机号码是有固定格式的，比如早期的手机号码都是"13"字段开头的，后来随着移动业务的发展又增加了"18"字段开头的，下面就以这两个字段为例进行测试，具体代码如下：

【代码 8-8】（详见源代码目录 ch08-react-form-tel-format.html 文件）

```
01  <!DOCTYPE html>
02  <html>
03  <head>
04      <meta charset="UTF-8" />
05      <title>React Form - React Form</title>
06      <script src="https://unpkg.com/react@16/umd/react.development.js"></script>
07      <script src="https://unpkg.com/react-dom@16/umd/react-dom.development.js"></script>
08      <!-- Don't use this in production: -->
09      <script src="https://unpkg.com/babel-standalone@6.15.0/babel.min.js"></script>
10  </head>
11  <body>
12  <!-- 添加文档主体内容 -->
13  <div id='id-div-react'></div>
14  <script type="text/babel">
15      // TODO: get div
16      var divReact = document.getElementById('id-div-react');
17      // TODO: define ES6 Class Component
18      class FormComp extends React.Component {
19          // TODO: Component Constructor
```

```
20        constructor(props) {
21            super(props);
22            this.state = {
23                tel: '',
24                info: '138****8000'
25            };
26            this.handleTelChange = this.handleTelChange.bind(this);
27        }
28        // TODO: handle event
29        handleTelChange(event) {
30            // TODO: define RegExp of Mobile Phone Number
31            let telReg = /13[0-9]\d{8}|18[56789]\d{8}/g;
32            let targetTel = event.target.value;
33            let finalTel = targetTel;
34            if(targetTel.length < 11) {
35                this.setState({info: 'pls go on...'});
36            } else if(targetTel.length == 11) {
37                if(telReg.test(targetTel)) {
38                    this.setState({info: 'right number.'});
39                    finalTel = targetTel;
40                } else {
41                    this.setState({info: 'wrong number.'});
42                    finalTel = "";
43                }
44            } else {
45                finalTel = targetTel.substr(0, 11);
46            }
47            this.setState({tel: finalTel});
48        }
49        // TODO: render form
50        render() {
51            return (
52                <form>
53                    <label>Mobile Phone(+86):</label><br/>
54                    <input type="text" name="tel"
55                        value={this.state.tel}
56                        onChange={this.handleTelChange} />
57                    <input type="text" name="info" className="css-input-info"
58                        value={this.state.info} readOnly/><br/><br/>
59                    <input type="submit" value="Submit" />
60                </form>
61            );
```

```
62          }
63      }
64      // TODO: React JSX
65      const reactSpan = (
66          <span>
67              <h3>React Form - Mobile Phone Length</h3>
68              <FormComp />
69          </span>
70      );
71      // TODO: React render
72      ReactDOM.render(reactSpan, divReact);
73 </script>
74 </body>
75 </html>
```

关于【代码8-8】的说明：

- 这段代码是在【代码8-7】的基础上改进而成的，主要是在原有手机号码长度校验的基础上，又增加了格式的校验（仅针对"13"字段和"18"字段）。
- 第31行代码定义了一个正则表达式（telReg），用于校验"13"字段和"18"字段开头的手机号码。
- 第37~43行代码通过if...else...条件语句校验用户输入手机号码的格式，具体内容如下：
  - 在第37行代码if条件语句中，先通过调用RegExp对象（telReg）的test()方法校验用户输入的手机号码是否符合要求。
  - 然后根据if条件语句的判断结果，选择第38行或第41行代码通过setState()方法将提示信息更新到state状态属性（info），并同步更新到第57~58行代码定义的只读文本输入框。

测试网页，页面效果如图8.12、图8.13和图8.14所示。

图8.12 校验手机号码格式（一）

第 8 章　React 表单

图 8.13　校验手机号码格式（二）

图 8.14　校验手机号码格式（三）

图 8.12 和图 8.13 中的提示信息（right number）表明号码格式正确，而图 8.14 中的提示信息（wrong number）表明号码格式不符合要求。

8.5　格式化序列号

在本节中，我们再实现一个格式化序列号输入应用场景的代码实例。相信大多数读者都有过安装 Windows 操作系统的经历，在安装正版操作系统的过程中都会遇到一个输入"CD-KEY"的界面，并提示用户在安装光盘的封面上查找。这个"CD-KEY"也称为序列号，格式统一为"XXXXX-XXXXX-XXXXX-XXXXX-XXXXX"。细心的读者会注意到，序列号中的短横线"-"是不需要用户输入的，当用户每输入完 5 个一组的字母或数字后，短横线"-"符号是自动添加的。

下面，就看一下如何通过 React 文本输入框"受控组件"，模拟实现一个格式化序列号（简

187

化形式：XXX-XXX-XXX）的代码实例，具体如下：

【代码 8-9】（详见源代码目录 ch08-react-form-sn-format.html 文件）

```html
01  <!DOCTYPE html>
02  <html>
03  <head>
04      <meta charset="UTF-8" />
05      <title>React Form - React Form</title>
06      <script src="https://unpkg.com/react@16/umd/react.development.js"></script>
07      <script src="https://unpkg.com/react-dom@16/umd/react-dom.development.js"></script>
08      <!-- Don't use this in production: -->
09      <script src="https://unpkg.com/babel-standalone@6.15.0/babel.min.js"></script>
10  </head>
11  <body>
12  <!-- 添加文档主体内容 -->
13  <div id='id-div-react'></div>
14  <script type="text/babel">
15      // TODO: get div
16      var divReact = document.getElementById('id-div-react');
17      // TODO: define ES6 Class Component
18      class FormComp extends React.Component {
19          // TODO: Component Constructor
20          constructor(props) {
21              super(props);
22              this.state = {
23                  sn: ''
24              };
25              this.handleSNChange = this.handleSNChange.bind(this);
26          }
27          // TODO: handle event
28          handleSNChange(event) {
29              let targetSN = event.target.value;
30              let finalSN = targetSN;
31              if(targetSN.length > 11) {
32                  finalSN = targetSN.substr(0, 11);
33              } else {
34                  if(targetSN.length == 3) {
35                      finalSN = targetSN + "-";
36                  } else if(targetSN.length == 7) {
37                      finalSN = targetSN + "-";
```

```
38                } else {}
39            }
40            this.setState({sn: finalSN});
41        }
42        // TODO: render form
43        render() {
44            return (
45                <form>
46                    <label>Serial Number:</label><br/>
47                    <input type="text" name="sn"
48                        value={this.state.sn}
49                        onChange={this.handleSNChange} /><br/><br/>
50                    <input type="submit" value="Submit" />
51                </form>
52            );
53        }
54    }
55    // TODO: React JSX
56    const reactSpan = (
57        <span>
58            <h3>React Form - Serial Number Format</h3>
59            <FormComp />
60        </span>
61    );
62    // TODO: React render
63    ReactDOM.render(reactSpan, divReact);
64  </script>
65  </body>
66  </html>
```

关于【代码 8-9】的说明：

- 这段代码的核心部分就是第 28～41 行代码实现的 "onChange" 事件处理方法（handleSNChange），具体内容如下：
  - 第 29 行代码先通过 "event" 对象获取了用户输入的内容。
  - 第 31～39 行代码通过 if...else...条件语句检测用户输入序列号的长度（字母、数字和连接符 "-" 共计 11 位）。如果长度等于 11 位，就会强制禁止用户输入（见第 32 行代码的 JavaScript 脚本处理方式）。
  - 第 34～38 行代码是第二个的 if...else...条件语句，用于监控用户输入序列号的长度。当长度等于 3 和 7 位时，自动添加连接符 "-"。

测试网页，页面效果如图 8.15 和图 8.16 所示。当用户在文本输入框中输入序列号的过程中，连接符 "-" 会根据长度自动添加。

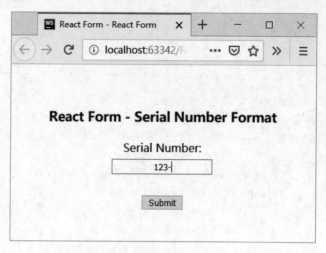

图 8.15　格式化序列号（一）

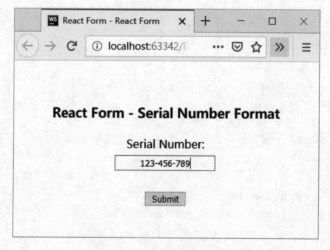

图 8.16　格式化序列号（二）

8.6　文本域关键字

在网页表单（Form）设计中，除了文本输入框这类页面元素外，类似的还有文本域页面元素。一般地，文本输入框仅支持单行文本，而文本域则可以支持多行文本。所以，文本域适用于输入大段的文本内容。

在本节中，我们设计实现一个可以自动识别关键字、并强制将关键字由小写字母转换为大写字母的 React 文本域 "受控组件"，具体代码如下：

【代码 8-10】（详见源代码目录 ch08-react-form-textarea-keyword.html 文件）

```
01  <!DOCTYPE html>
```

```
02  <html>
03  <head>
04      <meta charset="UTF-8" />
05      <title>React Form - React Form</title>
06      <script src="https://unpkg.com/react@16/umd/react.development.js"></script>
07      <script src="https://unpkg.com/react-dom@16/umd/react-dom.development.js"></script>
08      <!-- Don't use this in production: -->
09      <script src="https://unpkg.com/babel-standalone@6.15.0/babel.min.js"></script>
10  </head>
11  <body>
12  <!-- 添加文档主体内容 -->
13  <div id='id-div-react'></div>
14  <script type="text/babel">
15      // TODO: get div
16      var divReact = document.getElementById('id-div-react');
17      // TODO: define ES6 Class Component
18      class FormComp extends React.Component {
19          // TODO: Component Constructor
20          constructor(props) {
21              super(props);
22              this.state = {
23                  keyword: ''
24              };
25              this.handleArticleChange = this.handleArticleChange.bind(this);
26          }
27          // TODO: handle event
28          handleArticleChange(event) {
29              let target_keyword = event.target.value;
30              let markdown_keyword = target_keyword.replace(/king/g, "KING");
31              this.setState({keyword: markdown_keyword});
32          }
33          // TODO: render form
34          render() {
35              return (
36                  <form>
37                      <label>Article(keyword:king):</label><br/>
38                      <textarea type="text" name="article"
39                              value={this.state.keyword}
40                              onChange={this.handleArticleChange} /><br/><br/>
```

```
41                    <input type="submit" value="Submit" />
42                </form>
43            );
44        }
45    }
46    // TODO: React JSX
47    const reactSpan = (
48        <span>
49            <h3>React Form - Textarea Keyword</h3>
50            <FormComp />
51        </span>
52    );
53    // TODO: React render
54    ReactDOM.render(reactSpan, divReact);
55 </script>
56 </body>
57 </html>
```

关于【代码 8-10】的说明：

- 第 34～44 行代码通过 render()方法渲染了一个表单，第 38～40 行代码通过<textarea>标签元素定义了一个文本域"受控组件"。注意，"value"属性绑定了 state 状态属性（keyword），"onChange"事件处理方法定义为（handleArticleChange），形式与文本输入框"受控组件"相同。

- 第 20～26 行代码定义的构造方法中，第 23 行代码定义了一个 state 状态属性（keyword），绑定给第 39 行代码定义的文本域中的"value"属性；第 25 行代码将第 40 行代码定义的"onChange"事件处理方法（handleArticleChange）通过 bind()方法进行了绑定。

- 这段代码的核心部分是第 28～32 行代码实现的"onChange"事件处理方法（handleArticleChange），具体内容如下：
  ➢ 第 29 行代码先通过"event"对象获取了用户输入的内容。
  ➢ 第 30 行代码通过正则表达式查找匹配的模式字符串（/king/g），然后通过 replace()方法将查找到的字符串强制转换为大写字母。
  ➢ 第 31 行代码通过 setState()方法将操作结果更新到 state 状态属性（keyword），并同步更新到第 38～40 行代码定义的文本域。

测试网页，页面效果如图 8.17 所示。如图中的标识所示，React 文本域"受控组件"完成了对两处关键字"king"的自动识别。不过【代码 8-10】并不完美，因为第 2 处所识别的关键字"king"实际是单词"kingdom"的一部分，这相当于是误操作了。

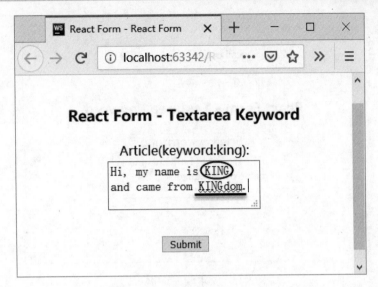

图 8.17 文本域关键字（一）

如何纠正这个错误呢？我们需要借助正则表达式，再结合条件语句，改进一下【代码 8-10】中的 "onChange" 事件处理方法（handleArticleChange），具体代码如下：

【代码 8-11】（详见源代码目录 ch08-react-form-textarea-keyword-m.html 文件）

```
01  // TODO: handle event
02  handleArticleChange(event) {
03      let regexp_keyword = /\sking\s/g;
04      let target_keyword = event.target.value;
05      let markdown_keyword = target_keyword;
06      if(regexp_keyword.test(target_keyword)) {
07          markdown_keyword = target_keyword.replace(regexp_keyword, " KING ");
08      }
09      this.setState({keyword: markdown_keyword});
10  }
```

关于【代码 8-11】的说明：

- 第 03 行代码定义了正则表达式模式字符串（/\sking\s/g），新增的模式字符串 "\s" 用于识别任意空白字符。
- 第 06~08 行代码通过 if 条件语句来判断检索模式字符串的结果，如果查找到了模式字符串，就通过 replace()方法进行关键字的替换。

测试网页，页面效果如图 8.18 所示。如图中的箭头所示，改进后的 React 文本域 "受控组件"可以区分关键字 "king" 和包含字符串 "king" 的单词了。

图 8.18 文本域关键字（二）

8.7 下拉列表受控组件

在本节中，我们再介绍 React 表单中如何实现比较常用的下拉列表"受控组件"，请看下面的代码实例：

【代码 8-12】（详见源代码目录 ch08-react-form-select.html 文件）

```
01  <!DOCTYPE html>
02  <html>
03  <head>
04      <meta charset="UTF-8" />
05      <title>React Form - React Form</title>
06      <script src="https://unpkg.com/react@16/umd/react.development.js"></script>
07      <script src="https://unpkg.com/react-dom@16/umd/react-dom.development.js"></script>
08      <!-- Don't use this in production: -->
09      <script src="https://unpkg.com/babel-standalone@6.15.0/babel.min.js"></script>
10  </head>
11  <body>
12  <!-- 添加文档主体内容 -->
13  <div id='id-div-react'></div>
14  <script type="text/babel">
15      // TODO: get div
16      var divReact = document.getElementById('id-div-react');
```

```
17      // TODO: define ES6 Class Component
18      class FormComp extends React.Component {
19         // TODO: Component Constructor
20         constructor(props) {
21            super(props);
22            this.state = {
23               selval: 'football'
24            };
25            this.handleSelChange = this.handleSelChange.bind(this);
26         }
27         // TODO: handle event
28         handleSelChange(event) {
29            let selVal = event.target.value;
30            console.log("You has selected '" + selVal + "'.");
31            this.setState({
32               selval: selVal
33            });
34         }
35         // TODO: render form
36         render() {
37            return (
38               <form>
39                  <label>
40                     Pls select your favorite sports:`
41                     <select value={this.state.selval} onChange={this.handleSelChange}>
42                        <option value="baseball">Baseball</option>
43                        <option value="football">Football</option>
44                        <option value="basketball">Basketball</option>
45                     </select>
46                  </label>
47                  <input type="submit" value="提交" />
48               </form>
49            );
50         }
51      }
52      // TODO: React JSX
53      const reactSpan = (
54         <span>
55            <h3>React Form - Select</h3>
56            <FormComp />
57         </span>
58      );
```

```
59        // TODO: React render
60        ReactDOM.render(reactSpan, divReact);
61    </script>
62 </body>
63 </html>
```

关于【代码 8-12】的说明：

- 第 36~50 行代码通过 render()方法渲染了一个表单，第 41~45 行代码通过<select>标签元素定义了一个下拉列表"受控组件"。注意，"value"属性绑定了 state 状态属性（selval），"onChange"事件处理方法定义为 handleSelChange。
- 第 20~26 行代码定义的构造方法中，第 23 行代码定义了一个 state 状态属性（selval）绑定给第 41 行代码定义的下拉列表中的"value"属性，并进行了初始化（'football'）；第 25 行代码将第 41 行代码定义的"onChange"事件处理方法（handleSelChange）通过 bind()方法进行了绑定。
- 第 28~34 行代码实现了"onChange"事件的处理方法（handleSelChange），具体内容如下：
  ➢ 第 29 行代码先通过"event"对象获取了用户所选择的下拉列表项的内容。
  ➢ 第 30 行代码通过浏览器控制台输出了获取的下拉列表项内容。
  ➢ 第 31~33 行代码通过 setState()方法将操作结果更新到 state 状态属性（selval），并同步更新到第 41~45 行代码定义的下拉列表。

测试网页，页面初始效果如图 8.19 所示。页面中下拉列表初始项为（Football），是由第 23 行代码定义的属性（selval）初始值所决定的。

图 8.19 下拉列表"受控组件"（一）

下面，我们可以自行选择一个下拉列表项，页面效果如图 8.20 所示。如图中的箭头所示，浏览器控制台中显示出了用户重新选择下拉列表项的操作记录。

第 8 章 React 表单

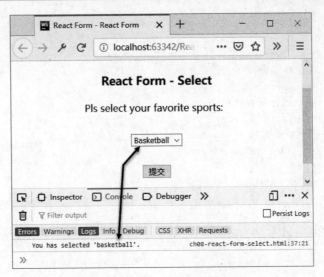

图 8.20 下拉列表"受控组件"(二)

8.8 处理多个输入

在 React 表单中，如果需要处理多个<input>标签元素，可以通过给每个元素添加各自"name"属性，并通过同一个事件处理方法进行处理。至于如何在同一个事件处理方法中区分不同的<input>标签元素，则是根据传递进来 Event 对象的"name"属性值进行操作的。

下面，我们介绍一个在 React 表单中如何实现处理多个<input>标签元素的代码实例，具体如下：

【代码 8-13】（详见源代码目录 ch08-react-form-multi-input.html 文件）

```
01  <!DOCTYPE html>
02  <html>
03  <head>
04      <meta charset="UTF-8" />
05      <title>React Form - React Form</title>
06      <script src="https://unpkg.com/react@16/umd/react.development.js"></script>
07      <script src="https://unpkg.com/react-dom@16/umd/react-dom.development.js"></script>
08      <!-- Don't use this in production: -->
09      <script src="https://unpkg.com/babel-standalone@6.15.0/babel.min.js"></script>
10  </head>
11  <body>
12  <!-- 添加文档主体内容 -->
13  <div id='id-div-react'></div>
14  <script type="text/babel">
15      // TODO: get div
```

```
16      var divReact = document.getElementById('id-div-react');
17      // TODO: define ES6 Class Component
18      class FormComp extends React.Component {
19          // TODO: Component Constructor
20          constructor(props) {
21              super(props);
22              this.state = {
23                  isOnOff: true,
24                  username: "king"
25              };
26              this.handleInputChange = this.handleInputChange.bind(this);
27          }
28          // TODO: handle event
29          handleInputChange(event) {
30              const target = event.target;
31              const value = target.type === 'checkbox' ? target.checked : target.value;
32              const name = target.name;
33              this.setState({
34                  [name]: value
35              });
36              console.log(name + " : " + value);
37          }
38          // TODO: render form
39          render() {
40              return (
41                  <form>
42                      <label>
43                          On/Off:  
44                          <input
45                              name="isOnOff"
46                              type="checkbox"
47                              checked={this.state.isOnOff}
48                              onChange={this.handleInputChange} />
49                      </label>
50                      <br/>
51                      <label>
52                          Username:  
53                          <input
54                              name="username"
55                              type="text"
56                              value={this.state.username}
57                              onChange={this.handleInputChange} />
58                      </label>
59                  </form>
60              );
61          }
62      }
63      // TODO: React JSX
64      const reactSpan = (
65          <span>
```

```
66              <h3>React Form - Multi Input</h3>
67              <FormComp />
68          </span>
69      );
70      // TODO: React render
71      ReactDOM.render(reactSpan, divReact);
72  </script>
73  </body>
74  </html>
```

关于【代码 8-13】的说明：

- 第 39~61 行代码通过 render()方法渲染了一个表单，其中定义了两个不同类型的 <input>标签元素，具体内容如下：
  - 第 44~48 行代码通过<input>标签元素定义了一个单选框（type=checkbox），"checked"属性绑定了 state 状态属性（isOnOff），"onChange"事件处理方法定义为（handleInputChange）。
  - 第 53~57 行代码通过<input>标签元素定义了一个文本输入框（type=text），"value"属性绑定了 state 状态属性（username），"onChange"事件处理方法同样定义为（handleInputChange）。
- 第 20~27 行代码定义了构造方法，具体内容如下：
  - 第 23 行代码定义了第 1 个 state 状态属性（isOnOff），绑定到第 47 行代码定义的下拉列表中的"checked"属性，并进行了初始化（true）。
  - 第 24 行代码定义了第 2 个 state 状态属性（username），绑定到第 56 行代码定义的下拉列表中的"value"属性，并进行了初始化（"king"）。
  - 第 26 行代码将"onChange"事件处理方法（handleInputChange）通过 bind()方法进行了绑定。
- 第 29~37 行代码实现了"onChange"事件的处理方法（handleInputChange），具体内容如下：
  - 第 30 行代码先通过"event"对象获取了用户操作的目标节点（target）。
  - 第 31 行代码通过三元条件表达式判断目标节点的类型（target.type），然后获取目标节点（target）对应的值（value）。需要注意一点，单选框的值与文本输入框的值，二者类型不同。
  - 第 32 行代码获取了目标节点元素的内容属性（name）。
  - 第 33~35 行代码通过 setState()方法将操作结果更新到 state 状态属性（通过 name 属性进行操作）。
  - 第 36 行代码通过浏览器控制台输出了获取的 name 属性和 value 值。

测试网页，页面效果如图 8.21 和图 8.22 所示。如图中的箭头所示，浏览器控制台中显示出了用户操作单选框和文本输入框的记录。

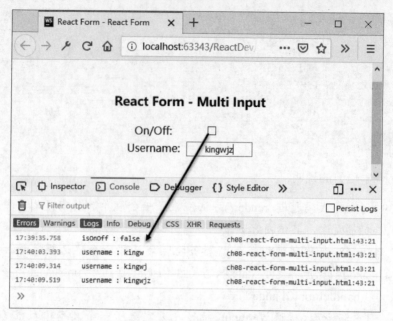

图 8.21　处理多个输入（一）

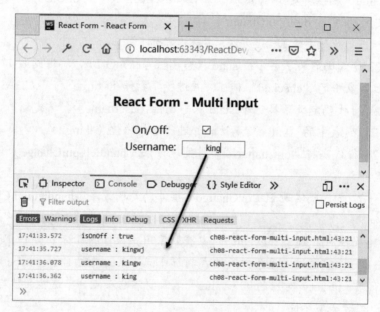

图 8.22　处理多个输入（二）

8.9　React 表单提交操作

前面介绍了很多关于 React 表单组件设计方面的内容，其实表单还由一个很主要的作用就是提交（Submit）操作。下面，我们介绍一个最基本的、关于 React 表单提交（Submit）操作的代码实例，具体如下：

【代码 8-14】（详见源代码目录 ch08-react-form-sumit.html 文件）

```html
01  <!DOCTYPE html>
02  <html>
03  <head>
04      <meta charset="UTF-8" />
05      <title>React Form - React Form</title>
06      <script src="https://unpkg.com/react@16/umd/react.development.js"></script>
07      <script src="https://unpkg.com/react-dom@16/umd/react-dom.development.js"></script>
08      <!-- Don't use this in production: -->
09      <script src="https://unpkg.com/babel-standalone@6.15.0/babel.min.js"></script>
10  </head>
11  <body>
12  <!-- 添加文档主体内容 -->
13  <div id='id-div-react'></div>
14  <script type="text/babel">
15      // TODO: get div
16      var divReact = document.getElementById('id-div-react');
17      // TODO: define ES6 Class Component
18      class FormComp extends React.Component {
19          // TODO: Component Constructor
20          constructor(props) {
21              super(props);
22              this.state = {
23                  username: "",
24                  age: "18",
25                  gender: "male"
26              };
27              this.handleInputChange = this.handleInputChange.bind(this);
28              this.handleSelChange = this.handleSelChange.bind(this);
29              this.handleSubmit = this.handleSubmit.bind(this);
30          }
31          // TODO: handle event
32          handleInputChange(event) {
33              const target = event.target;
34              const value = target.type === 'checkbox' ? target.checked : target.value;
35              const name = target.name;
36              this.setState({
37                  [name]: value
38              });
```

```
39              //console.log(name + " : " + value);
40          }
41          handleSelChange(event) {
42              let gender = event.target.value;
43              this.setState({
44                  gender: gender
45              });
46              //console.log("selected '" + gender + "'.");
47          }
48          handleSubmit(event) {
49              event.preventDefault();
50              console.log("Username : " + this.state.username);
51              console.log("Age : " + this.state.age);
52              console.log("Gender : " + this.state.gender);
53          }
54          // TODO: render form
55          render() {
56              return (
57                  <form onSubmit={this.handleSubmit}>
58                      <label>
59                          Username:  
60                          <input
61                              name="username"
62                              type="text"
63                              value={this.state.username}
64                              onChange={this.handleInputChange} />
65                      </label><br/>
66                      <label>
67                          Age:  
68                          <input
69                              name="age"
70                              type="number"
71                              value={this.state.age}
72                              onChange={this.handleInputChange} />
73                      </label><br/>
74                      <label>
75                          Gender:  
76      <select name="gender" value={this.state.gender} onChange={this.handleSelChange}>
77                          <option value="male">Male</option>
78                          <option value="female">Female</option>
79                      </select>
80                      </label><br/><br/>
```

```
81                <input type="submit" value="提交" />
82            </form>
83        );
84    }
85  }
86  // TODO: React JSX
87  const reactSpan = (
88    <span>
89        <h3>React Form - Submit</h3>
90        <FormComp />
91    </span>
92  );
93  // TODO: React render
94  ReactDOM.render(reactSpan, divReact);
95  </script>
96  </body>
97  </html>
```

关于【代码 8-14】的说明：

- 第 55~84 行代码通过 render()方法渲染了一个表单，其中定义了<input>标签元素、<select>下拉列表和提交（submit）按钮，具体内容如下：
  - 第 57 行代码通过<form>标签元素定义了一个表单，并定义了"onSubmit"事件处理方法定义为（handleSubmit）。
  - 第 60~64 行代码通过<input>标签元素定义了第一个文本输入框（type=text），"value"属性绑定了 state 状态属性（username），"onChange"事件处理方法定义为（handleInputChange）。
  - 第 68~72 行代码通过<input>标签元素定义了第二个数字型文本输入框（type=number），"value"属性绑定了 state 状态属性（age），"onChange"事件处理方法同样定义为（handleInputChange）。
  - 第 76~79 行代码通过<select>标签元素定义了一个下拉列表框，"value"属性绑定了 state 状态属性（gender），"onChange"事件处理方法定义为（handleSelChange）。
- 第 20~30 行代码定义了构造方法，具体内容如下：
  - 第 23 行代码定义了第 1 个 state 状态属性（username），绑定给第 63 行中代码定义的文本输入框中的"value"属性。
  - 第 24 行代码定义了第 2 个 state 状态属性（age），绑定给第 71 行中代码定义的数字型文本输入框中的"value"属性，并初始化为 18。
  - 第 25 行代码定义了第 3 个 state 状态属性（gender），绑定给第 76 行中代码定义的下拉列表框中的"value"属性。

➢ 第 27 行代码将"onChange"事件处理方法（handleInputChange）通过 bind()方法进行了绑定。
➢ 第 28 行代码将"onChange"事件处理方法（handleSelChange）通过 bind()方法进行了绑定。
➢ 第 29 行代码将"onSubmit"事件处理方法（handleSubmit）通过 bind()方法进行了绑定。
● 第 32~40 行代码实现了"onChange"事件的处理方法（handleInputChange），第 41~47 行代码实现了"onChange"事件的处理方法（handleSelChange），读者可以参考前面章节中关于这两个事件处理方法的介绍。
● 第 48~53 行代码实现了"onSubmit"事件的处理方法（handleSubmit），具体内容如下：
➢ 第 49 行代码先通过"event"对象调用 preventDefault()方法来阻止事件的默认行为。
➢ 第 50~52 行代码通过 state 状态属性（username、age、gender）在浏览器控制台中输出表单提交的内容。

测试网页，页面的初始效果如图 8.23 所示。

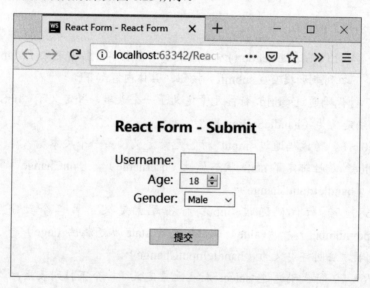

图 8.23 表单提交操作（一）

下面，我们可以尝试填写或修改一下表单项的内容，然后点击提交按钮进行测试，页面效果如图 8.24 所示。如图中的箭头所示，浏览器控制台中显示出了用户操作表单项后提交的表单内容。

第 8 章　React 表单

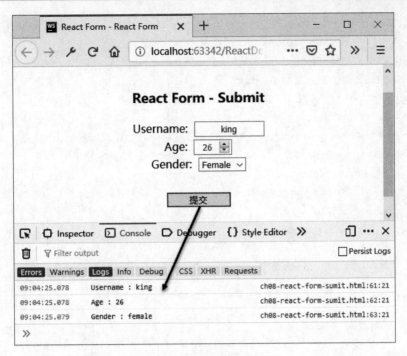

图 8.24　表单提交操作（二）

8.10　React 表单提交服务器

本节继续介绍关于 React 表单提交的内容。【代码 8-14】实现的表单提交操作，仅仅是在当前页面下完成的。在实际项目开发中，React 表单数据还是要提交到服务器端去处理，这就涉及 Web 前端与后台架构的知识。下面，我们介绍一个 React 表单提交到 Python 后端服务器的代码实例，具体如下：

【代码 8-15】（详见源代码目录 ch08-react-form-sumit-server.html 文件）

```
01  <!DOCTYPE html>
02  <html>
03  <head>
04      <meta charset="UTF-8" />
05      <title>React Form - React Form</title>
06      <script src="https://unpkg.com/react@16/umd/react.development.js"></script>
07      <script src="https://unpkg.com/react-dom@16/umd/react-dom.development.js"></script>
08      <!-- Don't use this in production: -->
09      <script src="https://unpkg.com/babel-standalone@6.15.0/babel.min.js"></script>
```

205

```
10    </head>
11    <body>
12    <!-- 添加文档主体内容 -->
13    <div id='id-div-react'></div>
14    <script type="text/babel">
15        // TODO: get div
16        var divReact = document.getElementById('id-div-react');
17        // TODO: define ES6 Class Component
18        class FormComp extends React.Component {
19            // TODO: Component Constructor
20            constructor(props) {
21                super(props);
22                this.state = {
23                    username: "",
24                    age: "18",
25                    gender: "male"
26                };
27                this.handleInputChange = this.handleInputChange.bind(this);
28                this.handleSelChange = this.handleSelChange.bind(this);
29            }
30            // TODO: handle event
31            handleInputChange(event) {
32                const target = event.target;
33                const value = target.type === 'checkbox' ? target.checked : target.value;
34                const name = target.name;
35                this.setState({
36                    [name]: value
37                });
38                //console.log(name + " : " + value);
39            }
40            handleSelChange(event) {
41                let gender = event.target.value;
42                this.setState({
43                    gender: gender
44                });
45                //console.log("selected '" + gender + "'.");
46            }
47            // TODO: render form
48            render() {
49                return (
50                    <form action="server.py" method="GET">
51                        <label>
```

```
52                        Username:  
53                        <input
54                            name="username"
55                            type="text"
56                            value={this.state.username}
57                            onChange={this.handleInputChange} />
58                    </label><br/>
59                    <label>
60                        Age:  
61                        <input
62                            name="age"
63                            type="number"
64                            value={this.state.age}
65                            onChange={this.handleInputChange} />
66                    </label><br/>
67                    <label>
68                        Gender:  
69     <select name="gender" value={this.state.gender} onChange={this.handleSelChange}>
70                            <option value="male">Male</option>
71                            <option value="female">Female</option>
72                        </select>
73                    </label><br/><br/>
74                    <input type="submit" value="提交" />
75                </form>
76            );
77        }
78    }
79    // TODO: React JSX
80    const reactSpan = (
81        <span>
82            <h3>React Form - Submit to Server</h3>
83            <FormComp />
84        </span>
85    );
86    // TODO: React render
87    ReactDOM.render(reactSpan, divReact);
88 </script>
89 </body>
90 </html>
```

关于【代码 8-15】的说明：

这段代码是在【代码 8-14】的基础上修改而成的，具体内容如下：

- 第 50 行代码通过<form>标签元素定义了一个表单，"action"属性定义了服务器端提交地址（"server.py"），"method"属性定义了提交方式（"GET"）。
- 第 74 行代码通过<input>标签元素定义了一个提交按钮（type=submit），不过并没有定义"onSubmit"事件处理方法。

下面，我们再看一下服务器端的处理方式，这里采用 Python 语言来编写服务器端代码，具体如下：

【代码 8-16】（详见源代码目录 server.py 文件）

```
01  #!D:\Python\Python37\python
02  # coding:utf-8
03  # CGI 处理模块
04  import cgi, cgitb
05  # 创建 FieldStorage 的实例化
06  form = cgi.FieldStorage()
07  # 获取数据
08  site_username = form.getvalue('username')
09  site_age = form.getvalue('age')
10  site_gender = form.getvalue('gender')
11  // TODO: output html page
12  print("Content-type:text/html")
13  print()
14  print("<html>")
15  print("<head>")
16  print("<meta charset=\"utf-8\">")
17  print("<title>React to Server</title>")
18  print("</head>")
19  print("<body>")
20  print("<h3>Username:%s</h3>" % (site_username))
21  print("<h3>Age:%s</h3>" % (site_age))
22  print("<h3>Gender:%s</h3>" % (site_gender))
23  print("</body>")
24  print("</html>")
```

关于【代码 8-16】的说明：

- 这段服务器代码通过 Python 语言获取了客户端（ch08-react-form-sumit-server.html 页面文件）提交的数据，然后输出到页面中进行显示。

下面使用 Firefox 浏览器运行测试客户端 HTML 网页，页面初始效果如图 8.25 所示。

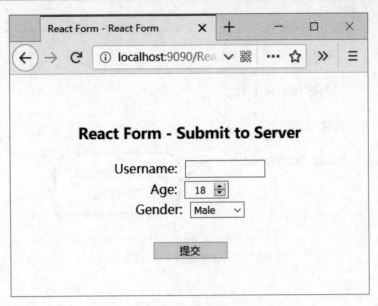

图 8.25　React 表单提交服务器（一）

我们尝试填写或修改表单项的内容，页面效果如图 8.26 所示。

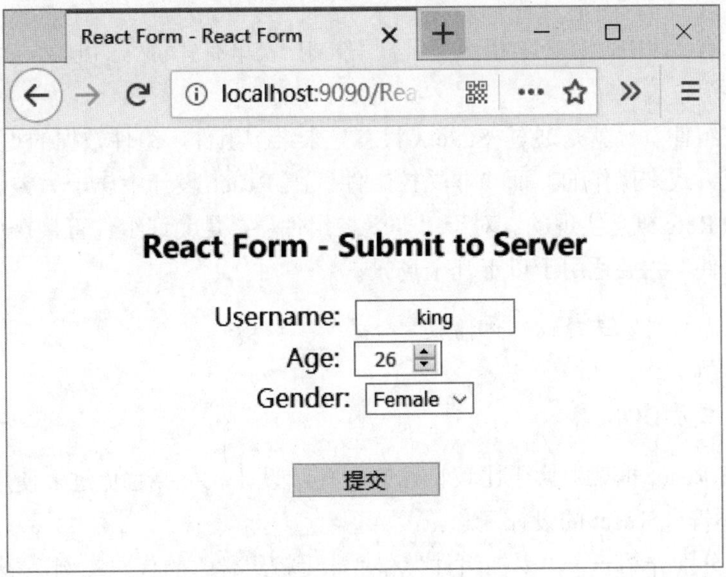

图 8.26　React 表单提交服务器（二）

然后，点击"提交"按钮尝试提交到服务器端进行测试，页面效果如图 8.27 所示。如图中的标识所示，页面中显示出了客户端 React 表单提交到服务器端的用户数据。

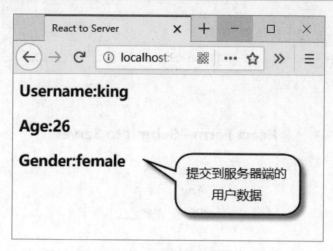

图 8.27　React 表单提交服务器（三）

8.11　受控组件与非受控组件

在 React 表单设计中，有一项相比于常规 HTML 表单有些特殊的概念，就是"受控组件"与"非受控组件"。

所谓"受控组件"，就是通过 State（状态）来维护组件，组件数据的更新是通过"自顶而下"的数据流方式来操作的。而"非受控组件"在 React 设计中属于另类，是通过与 State（状态）无关的 Refs 概念实现的。对于"非受控组件"，是允许组件直接操作实际 DOM 的。

"非受控组件"主要适用于以下几个场景：

- 管理焦点，文本选择或媒体播放。
- 触发强制动画。
- 集成第三方 DOM 库。

Refs 概念在 React 框架中属于比较特殊的一类，设计人员要避免过多使用 Refs 直接操作 DOM，因为这不符合 React 的设计理念。

下面，我们分别介绍一个"受控组件"和"非受控组件"的代码实例，帮助读者对"非受控组件"有更深入地了解。

首先，看一下"受控组件"的代码：

【代码 8-17】（详见源代码目录 ch08-react-refs-control.html 文件）

```
01  <!DOCTYPE html>
02  <html>
03  <head>
04      <meta charset="UTF-8"/>
05      <title>React Refs - Control Component</title>
```

```
06      <script src="https://unpkg.com/react@16/umd/react.development.js"></script>
07      <script src="https://unpkg.com/react-dom@16/umd/react-dom.development.js"></script>
08      <!-- Don't use this in production: -->
09      <script src="https://unpkg.com/babel-standalone@6.15.0/babel.min.js"></script>
10    </head>
11    <body>
12      <!-- 添加文档主体内容 -->
13      <div id='id-div-react'></div>
14      <script type="text/babel">
15        // TODO: get div
16        var divReact = document.getElementById('id-div-react');
17        // TODO: define ES6 Class Component
18        class ReactRefsComp extends React.Component {
19            constructor(props) {
20                super(props);
21                this.state = {
22                    username: '',
23                    output: ''
24                };
25                this.handleChange = this.handleChange.bind(this);
26                this.handleSubmit = this.handleSubmit.bind(this);
27            }
28            handleChange(e) {
29                this.setState({
30                    username: e.target.value
31                });
32            }
33            handleSubmit(e) {
34                e.preventDefault();
35                this.setState({
36                    output: 'Username is ' + this.state.username + '.'
37                });
38            }
39            render() {
40                return (
41                    <form onSubmit={this.handleSubmit}>
42                        <h3>React Refs - 受控组件</h3>
43                        <label>Input username:  
44                            <input
45                                type="text"
```

```
46                            value={this.state.username}
47                            onChange={this.handleChange}
48                        />
49                    </label>  
50                    <input type="submit" value="Submit" />
51                    <br/><br/>
52                    <p>{this.state.output}</p>
53                </form>
54            );
55        }
56    }
57    // TODO: React render
58    ReactDOM.render(<ReactRefsComp/>, divReact);
59 </script>
60 </body>
61 </html>
```

关于【代码 8-17】的说明：

- 这段代码实现的功能很简单，主要是设计了一个用于提交用户名的表单，其中用户名数据是通过 State（状态）属性（username）来维护的。

测试网页，页面效果如图 8.28 所示。如图中的箭头所示，用户名输入框（受控组件）中的信息成功提交到页面中显示了。

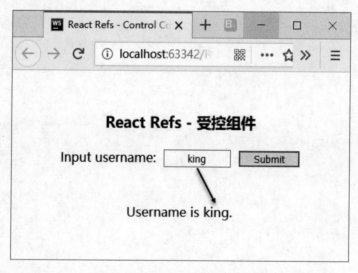

图 8.28　React 表单 "受控组件"

下面，再看一下 "非受控组件" 的代码：

【代码 8-18】（详见源代码目录 ch08-react-refs-uncontrol.html 文件）

```
01  <!DOCTYPE html>
```

```
02  <html>
03  <head>
04      <meta charset="UTF-8"/>
05      <title>React Refs - Uncontrol Component</title>
06      <script src="https://unpkg.com/react@16/umd/react.development.js"></script>
07      <script src="https://unpkg.com/react-dom@16/umd/react-dom.development.js"></script>
08      <!-- Don't use this in production: -->
09      <script src="https://unpkg.com/babel-standalone@6.15.0/babel.min.js"></script>
10  </head>
11  <body>
12  <!-- 添加文档主体内容 -->
13  <div id='id-div-react'></div>
14  <script type="text/babel">
15      // TODO: get div
16      var divReact = document.getElementById('id-div-react');
17      // TODO: define ES6 Class Component
18      class ReactRefsComp extends React.Component {
19          constructor(props) {
20              super(props);
21              this.state = {
22                  output: ''
23              };
24              this.handleSubmit = this.handleSubmit.bind(this);
25          }
26          handleSubmit(e) {
27              e.preventDefault();
28              this.setState({
29                  output: 'Username is ' + this.username.value + '.'
30              });
31          }
32          render() {
33              return (
34                  <form onSubmit={this.handleSubmit}>
35                      <h3>React Refs - 非受控组件</h3>
36                      <label>Input username:  
37                          <input
38                              type="text"
39                              ref={input => {this.username = input}}
40                          />
41                      </label>  
```

```
42                  <input type="submit" value="Submit" />
43                  <br/><br/>
44                  <p>{this.state.output}</p>
45              </form>
46          );
47      }
48  }
49  // TODO: React render
50  ReactDOM.render(<ReactRefsComp/>, divReact);
51  </script>
52  </body>
53  </html>
```

关于【代码8-18】的说明：

- 这段代码所实现的功能与【代码8-17】完全一致，区别就是实现过程采用了"非受控组件"的方式，具体说明如下：
  > 第37~40行代码定义的文本输入框（input）中，定义了一个"ref"属性（见第39行代码），该属性通过回调函数（箭头函数方式）定义了对 DOM 的引用（this.username），这样在组件的其他地方就可以直接操作该文本输入框了。第29行代码中，就是直接通过操作（this.username）获取了文本输入框中输入的用户名信息。

测试网页，页面效果如图 8.29 所示。如图中的箭头所示，用户名输入框（非受控组件）中的信息成功提交到页面中显示了，与图 8.28 所示的效果完全一致。

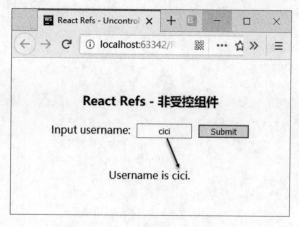

图 8.29　React 表单"非受控组件"

第 9 章

◀ 组合与继承 ▶

React 框架在组件代码重用方面，具有组合模式与继承模式。React 官方推荐设计人员在实际项目中尽量使用组合模式，而非继承模式来实现组件的代码重用。在本章中，我们将对这两种模式进行一些基本的阐述。

9.1 组合与继承概述

任何一门现代的高级编程语言，都会将代码重用作为重中之重的研究方向。比如，我们熟知的 C++和 Java 语言，就设计了类（Class）和类的继承概念，可以完美地支持代码重用功能。

React 框架是基于 ES6 规范标准设计的，自然也支持继承关系，通过 extends 关键字就可以实现。这里，在定义 React 类组件时所使用的代码（extends React.Component），就是继承操作。

那为什么 React 官方推荐设计人员使用组合模式呢？其实，并不是 React 框架不支持继承模式，关键是这两种模式用在什么地方。无论是实现比较抽象的类（类似 React 底层代码），还是推荐继承模式的，都是基于 ES6 的支持。但如果是现实中比较复杂的业务组件，考虑到代码质量问题，建议还是要通过组合模式来实现，这样可以避免设计人员在研究子组件时还要被迫去研究父组件。

9.2 定义组件容器

在很多情况下，设计人员在定义一些组件时，是无法提前考虑周全子组件的具体内容的。比如，在定义一个对话框（Dialog）通用组件时，就会遇到类似的情况。因此，React 建议先定义一个对话框通用容器（Box）组件，然后使用一个特殊的 Prop 属性（children）将未来的子组件传递到该容器中渲染，从而实现组件组合模式的设计。

下面，我们就先设计实现一个对话框容器组件，具体代码如下：

【代码 9-1】（详见源代码目录 ch09-react-compose-dialogbox.html 文件）

```
01  <script type="text/babel">
02      // TODO: base function component
03      function MyDialogBox(props) {
04          return (
05              <div className={'dialogBorder-' + props.color}>
06                  {props.children}
07              </div>
08          );
09      }
10  </script>
```

关于【代码 9-1】的说明：

- 第 03～09 行代码定义了一个对话框函数组件（MyDialogBox）容器，具体内容如下：
  - 第 03 行代码在定义函数组件时，添加了 prop 属性参数。
  - 第 04～08 行代码通过 return 方法返回了一个 <div> 标签容器。
  - 第 05 行代码通过 prop 属性参数传递进来对话框边框颜色属性（color）。
  - 第 06 行代码通过一个特殊的 prop 属性（children）传递进来对话框子组件的内容（此时是未定义的未知内容）。

读者已经注意到【代码 9-1】定义的对话框组件仅仅是一个容器，是没有实际内容展现的。那如何实现 React 组合模式呢？请读者继续往下阅读。

9.3 定义子组件

本节我们继续介绍前面章节的内容，设计一个对话框子组件（WelcomeDialog）。该子组件与【代码 9-1】定义的组件容器（MyDialogBox）构成组合模式（实际是包含关系），具体代码如下：

【代码 9-2】（详见源代码目录 ch09-react-compose-dialogbox-welcome.html 文件）

```
01  <!DOCTYPE html>
02  <html>
03  <head>
04      <meta charset="UTF-8" />
05      <title>React Form - React Compose</title>
06      <script src="https://unpkg.com/react@16/umd/react.development.js"></script>
07      <script src="https://unpkg.com/react-dom@16/umd/react-dom.development.js"></script>
```

```
08      <!-- Don't use this in production: -->
09      <script src="https://unpkg.com/babel-standalone@6.15.0/babel.min.js"></script>
10    </head>
11    <body>
12      <!-- 添加文档主体内容 -->
13      <div id='id-div-react'></div>
14      <script type="text/babel">
15        // TODO: get div
16        var divReact = document.getElementById('id-div-react');
17        // TODO: base function component
18        function MyDialogBox(props) {
19          return (
20            <div className={'dialogBorder-' + props.color}>
21              {props.children}
22            </div>
23          );
24        }
25        // TODO: sub function component
26        function WelcomeDialog() {
27          return (
28            <MyDialogBox color="gray">
29              <h3 className="dialog-title">
30                Welcome
31              </h3>
32              <p className="dialog-message">
33                Thanks for your visiting our website!
34              </p>
35            </MyDialogBox>
36          );
37        }
38        // TODO: React render
39        ReactDOM.render(
40          <WelcomeDialog />,
41          divReact
42        );
43      </script>
44    </body>
45  </html>
```

关于【代码9-2】的说明：

- 第 18～24 行代码定义了一个对话框函数组件（MyDialogBox）容器，读者可以参考【代码9-1】的内容。

- 第26~37行代码定义了一个对话框子组件（WelcomeDialog），其中第28~35行代码引用了第18~24行代码定义的对话框容器,该容器内的第29~34行代码定义了一些具体内容。
- 而正是第29~34行代码定义的具体内容，将通过prop参数的属性（children）传递给对话框函数组件容器，并最终渲染到页面中进行显示。

测试网页的效果如图9.1所示。如图中的标识所示，页面中显示了对话框组件（MyDialogBox）容器与对话框子组件（WelcomeDialog）组合模式的对话框组件。

图9.1 React 对话框组合

使用React组件组合模式的优势就是,可以基于一个组件容器和不同的子组件,构建出多种组件应用。下面,我们将【代码9-3】略作改进,并借助条件渲染方式实现一个多类型对话框组件应用,具体代码如下：

【代码9-3】（详见源代码目录 ch09-react-compose-dialogbox-multi.html 文件）

```
01    <!DOCTYPE html>
02    <html>
03    <head>
04        <meta charset="UTF-8" />
05        <title>React Form - React Compose</title>
06        <script src="https://unpkg.com/react@16/umd/react.development.js"></script>
07        <script src="https://unpkg.com/react-dom@16/umd/react-dom.development.js"></script>
08        <!-- Don't use this in production: -->
09        <script src="https://unpkg.com/babel-standalone@6.15.0/babel.min.js"></script>
```

```html
10  </head>
11  <body>
12  <!-- 添加文档主体内容 -->
13  <div id='id-div-react'></div>
14  <script type="text/babel">
15      // TODO: get div
16      var divReact = document.getElementById('id-div-react');
17      // TODO: base function component
18      function MyDialogBox(props) {
19        return (
20          <div className={'dialogBorder-' + props.color}>
21            {props.children}
22          </div>
23        );
24      }
25      // TODO: sub function component
26      function WelcomeDialog() {
27        return (
28          <MyDialogBox color="gray">
29            <h3 className="dialog-title">
30              Welcome
31            </h3>
32            <p className="dialog-message">
33              Thanks for your visiting our website!
34            </p>
35          </MyDialogBox>
36        );
37      }
38      // TODO: sub function component
39      function InfoDialog() {
40        return (
41          <MyDialogBox color="gray">
42            <h3 className="dialog-title">
43              Info
44            </h3>
45            <p className="dialog-info">
46              Attention: Pls read this information in detail!
47            </p>
48          </MyDialogBox>
49        );
50      }
51      // TODO: outer function component
52      function MyDialog(props) {
```

```
53        const dType = props.dlgType;
54        if(dType) {
55            console.log("Dialog Type: WelcomeDialog.");
56            return <WelcomeDialog />;
57        } else {
58            console.log("Dialog Type: InfoDialog.");
59            return <InfoDialog />;
60        }
61    }
62    // TODO: React render
63    ReactDOM.render(
64        <MyDialog dlgType={true} />,
65        divReact
66    );
67 </script>
68 </body>
69 </html>
```

关于【代码9-3】的说明：

- 第18~24行代码定义了一个对话框函数组件（MyDialogBox）容器，读者可以参考【代码9-1】的内容。
- 第26~37行代码和第39~50行代码分别定义了一个对话框子组件（WelcomeDialog和InfoDialog），设计目标是让这两个对话框子组件共享对话框函数组件容器。
- 第52~61行代码定义了一个函数组件（MyDialog），通过判断第64行代码定义的条件来选择渲染相应的对话框子组件（WelcomeDialog和InfoDialog）。

测试网页的效果如图9.2所示。由于第64行代码定义的初始条件为"true"，因此函数组件（MyDialog）通过条件渲染方式选择了对话框子组件（WelcomeDialog）。

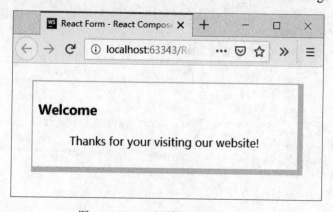

图9.2 React对话框子组件（一）

我们可以尝试将第 64 行代码定义的条件改为 "false"，然后再使用 Firefox 浏览器运行测试，效果如图 9.3 所示。此时函数组件通过条件渲染方式选择了对话框子组件。

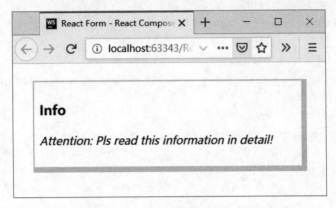

图 9.3　React 对话框子组件（二）

9.4　自定义 Props 属性

React 组件组合模式除了可以使用 prop 参数属性（children）之外，还支持自行约定 prop 参数属性。这种自行约定方式摆脱了 children 的约束，使用起来更加灵活，只需要将所需内容传入 prop 参数，并使用相对应的属性，具体代码如下：

【代码 9-4】（详见源代码目录 ch09-react-compose-panels.html 文件）

```
01  <!DOCTYPE html>
02  <html>
03  <head>
04      <meta charset="UTF-8" />
05      <title>React Form - React Compose</title>
06      <script src="https://unpkg.com/react@16/umd/react.development.js"></script>
07      <script src="https://unpkg.com/react-dom@16/umd/react-dom.development.js"></script>
08      <!-- Don't use this in production: -->
09      <script src="https://unpkg.com/babel-standalone@6.15.0/babel.min.js"></script>
10  </head>
11  <body>
12  <!-- 添加文档主体内容 -->
13  <div id='id-div-react'></div>
14  <script type="text/babel">
15      // TODO: get div
```

```
16      var divReact = document.getElementById('id-div-react');
17      // TODO: function component
18      function HeaderPanel() {
19          return <div className="css-header">Header Panel</div>;
20      }
21      function LeftPanel() {
22          return <div className="css-left">Left Panel</div>;
23      }
24      function RightPanel() {
25          return <div className="css-right">Right Panel</div>;
26      }
27      function MainPanel(props) {
28          return (
29              <div className="main-panel">
30                  <div className="main-panel-header">
31                      {props.header}
32                  </div>
33                  <div className="main-panel-left">
34                      {props.left}
35                  </div>
36                  <div className="main-panel-right">
37                      {props.right}
38                  </div>
39              </div>
40          );
41      }
42      // TODO: App component
43      function App() {
44          return (
45              <MainPanel
46                  header={
47                      <HeaderPanel />
48                  }
49                  left={
50                      <LeftPanel />
51                  }
52                  right={
53                      <RightPanel />
54                  } />
55          );
56      }
57      // TODO: React render
58      ReactDOM.render(
```

```
59          <App />,
60          divReact
61      );
62  </script>
63  </body>
64  </html>
```

关于【代码 9-4】的说明：

- 第 18 ~ 20 行代码、第 21 ~ 23 行代码和第 24 ~ 26 行代码定义了三个面板（Panel）组件（HeaderPanel、LeftPanel 和 RightPanel）。按照三个面板的字面含义，我们的目标是将这三个面板组成一个"头部、左侧栏和右侧栏"的页面布局。
- 第 27 ~ 41 行代码定义了一个主面板容器（MainPanel），通过 props 参数属性（header、left 和 right）将内容渲染到页面中。
- 第 27 ~ 41 行代码定义了一个组件入口（App），第 45 ~ 54 行代码引入了主面板容器（MainPanel），并通过添加属性（header、left 和 right）的方式引入上面定义的三个面板（HeaderPanel、LeftPanel 和 RightPanel）。

测试网页的效果如图 9.4 所示。页面中显示出了由"头部面板、左侧栏面板和右侧栏面板"组成的页面布局。

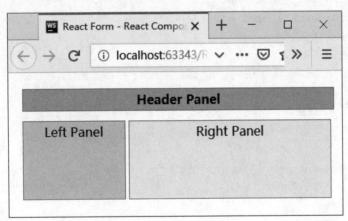

图 9.4 自定义 Props 属性

9.5 特例关系组合

React 组件还支持通过一种特例关系的组合方式来实现，此时这些组件可以被看作是其他组件的特殊实例。在 React 组件设计中，要实现特例关系组合可以通过定制 props 来渲染完成，请看下面的代码实例：

【代码 9-5】（详见源代码目录 ch09-react-compose-dialogbox-special.html 文件）

```html
01  <!DOCTYPE html>
02  <html>
03  <head>
04      <meta charset="UTF-8" />
05      <title>React Form - React Compose</title>
06      <script src="https://unpkg.com/react@16/umd/react.development.js"></script>
07      <script src="https://unpkg.com/react-dom@16/umd/react-dom.development.js"></script>
08      <!-- Don't use this in production: -->
09      <script src="https://unpkg.com/babel-standalone@6.15.0/babel.min.js"></script>
10  </head>
11  <body>
12  <!-- 添加文档主体内容 -->
13  <div id='id-div-react'></div>
14  <script type="text/babel">
15      // TODO: get div
16      var divReact = document.getElementById('id-div-react');
17      // TODO: base function component
18      function DialogBox(props) {
19          return (
20              <div className={'dialogBorder-' + props.color}>
21                  {props.children}
22              </div>
23          );
24      }
25      // TODO: universal function component
26      function Dialog(props) {
27          return (
28              <DialogBox color="gray">
29                  <h3 className="dialog-title">
30                      {props.title}
31                  </h3>
32                  <p className="dialog-message">
33                      {props.message}
34                  </p>
35              </DialogBox>
36          );
37      }
38      // TODO: special function component
39      function WelcomeDialog() {
```

```
40          return (
41              <Dialog
42                  title="Welcome"
43                  message="Thanks for your visiting our website!" />
44          );
45      }
46      // TODO: React render
47      ReactDOM.render(
48          <WelcomeDialog />,
49          divReact
50      );
51  </script>
52  </body>
53  </html>
```

关于【代码 9-5】的说明：

- 第 39~45 行代码定义了一个对话框组件（WelcomeDialog），而该组件可以看作是第 26~37 行代码定义的对话框组件（Dialog）的特殊实例。

测试网页的效果如图 9.5 所示。通过特例关系组合实现的组件与包含关系实现的组件是一致的，因此也可以将特例关系当作是一种特殊的包含关系。

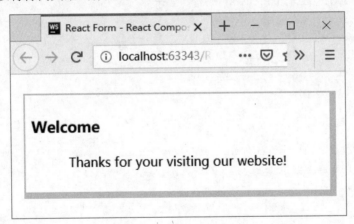

图 9.5 特殊关系组合

9.6 类组合方式确认框

前面代码实例中组合方式均是通过函数组件实现的，其实使用类（class）组件的组合方式也是可以实现的。本节，我们使用类（class）组件的组合方式实现一个简单的确认对话框，请看下面的代码实例：

【代码 9-6】（详见源代码目录 ch09-react-compose-dialogbox-confirm.html 文件）

```html
01  <!DOCTYPE html>
02  <html>
03  <head>
04      <meta charset="UTF-8" />
05      <title>React Form - React Compose</title>
06      <script src="https://unpkg.com/react@16/umd/react.development.js"></script>
07      <script src="https://unpkg.com/react-dom@16/umd/react-dom.development.js"></script>
08      <!-- Don't use this in production: -->
09      <script src="https://unpkg.com/babel-standalone@6.15.0/babel.min.js"></script>
10  </head>
11  <body>
12  <!-- 添加文档主体内容 -->
13  <div id='id-div-react'></div>
14  <script type="text/babel">
15      // TODO: get div
16      var divReact = document.getElementById('id-div-react');
17      // TODO: base function component
18      function DialogBox(props) {
19          return (
20              <div className={'dialogBorder-' + props.color}>
21                  {props.children}
22              </div>
23          );
24      }
25      // TODO: universal function component
26      function Dialog(props) {
27          return (
28              <DialogBox color="gray">
29                  <h3 className="dialog-title">
30                      {props.title}
31                  </h3>
32                  <p className="dialog-message">
33                      {props.message}
34                  </p>
35                  {props.children}
36              </DialogBox>
37          );
38      }
39      // TODO: special class component
```

```
40    class ConfirmDialog extends React.Component {
41        // TODO: Component Constructor
42        constructor(props) {
43            super(props);
44            this.handleConfirmClick = this.handleConfirmClick.bind(this);
45            this.handleCancelClick = this.handleCancelClick.bind(this);
46        }
47        // TODO: handle event
48        handleConfirmClick(event) {
49            console.log("you have selected 'Confirm' button.");
50        }
51        handleCancelClick(event) {
52            console.log("you have selected 'Cancel' button.");
53        }
54        // TODO: render Dialog
55        render() {
56            return (
57                <Dialog
58                    title="Confirm"
59                    message="Pls confirm your option!">
60                    <button onClick={this.handleConfirmClick}>Confirm</button>
61                    <button onClick={this.handleCancelClick}>Cancel</button>
62                </Dialog>
63            );
64        }
65    }
66    // TODO: React render
67    ReactDOM.render(
68        <ConfirmDialog />,
69        divReact
70    );
71 </script>
72 </body>
73 </html>
```

关于【代码 9-6】的说明：

- 【代码 9-6】沿用了【代码 9-5】的基础架构，对话框组件容器直接复制过来，主要变化就是第 40～65 行代码定义了一个类（class）对话框组件（ConfirmDialog），模拟实现了一个确认对话框，具体内容如下：
  > 第 58～59 行代码定义了两个属性（title 和 message），将通过第 26～38 行代码

定义的对话框组件（Dialog）的 props 参数属性（title 和 message）传递给第 30 行和第 33 行代码进行渲染。
➢ 第 60~61 行代码定义了两个按钮（Confirm 和 Cancel），同样通过对话框组件（Dialog）的 props 参数属性（children）传递给第 35 行代码进行渲染。

测试网页的效果如图 9.6 和图 9.7 所示。如图中的箭头所示，当我们分别点击"Confirm"和"Cancel"按钮后，浏览器控制中显示了第 49 行和第 52 行代码定义的调试信息，说明该代码实例模拟了确认对话框的功能。

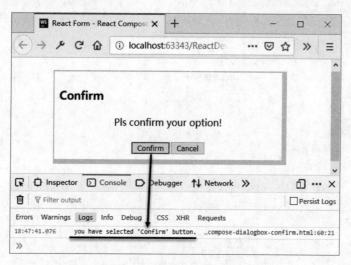

图 9.6　类组合方式实现确认对话框（一）

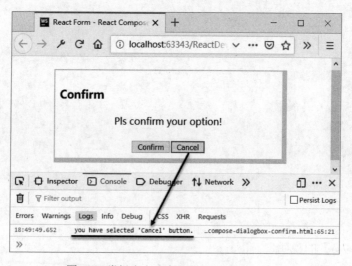

图 9.7　类组合方式实现确认对话框（二）

第 10 章

◀ 状态提升 ▶

在 React 框架中,当多个组件需要反映相同的变化数据,这时建议将共享状态提升到最近的共同父组件中去。一般我们将该操作称为"状态提升"。

10.1 设计构想

在前文关于"事件处理"的部分,我们曾经设计实现了一个"水温监控控件"的应用。这个控件实现了根据用户输入的水温,自动判断该温度具体是一个什么级别(Level)的功能。

在本章中,我们尝试将该应用在"状态提升"的功能方向上进行修改,可以同时满足不同的温度度量标准(摄氏温度和华氏温度)监控与同步,实现将共享状态(水温)提升到父一级的操作。

因此,我们可以将"水温监控控件"应用中,用于判断水温级别的函数组件(WaterTempLevel)提取出来,具体代码如下:

【代码 10-1】(详见源代码目录 ch10-react-water-temp-level.html 文件)

```
01  <script type="text/babel">
02      // TODO: define water temperature func Component
03      function WaterTempLevel(props) {
04          if (props.wlevel <= 0) {
05              return <p>This is ice water.</p>;
06          } else if((props.wlevel > 0) && (props.wlevel <= 20)) {
07              return <p>This is cold water.</p>;
08          } else if((props.wlevel > 20) && (props.wlevel <= 38)) {
09              return <p>This is warm water.</p>;
10          } else if((props.wlevel > 38) && (props.wlevel < 100)) {
11              return <p>This is hot water.</p>;
12          } else if(props.wlevel >= 100) {
13              return <p>This is boiling water.</p>;
14          } else {
15              return <p>This is ... water.</p>;
```

```
16          }
17      }
18  </script>
```

关于【代码 10-1】的说明：

- 这段代码主要就是通过 if-else 条件语句来判断用户输入水温的级别，根据不同的级别返回对应的渲染文本。

10.2 实现水温监控功能

在本节中，我们继续实现水温监控功能。首先，要设计一个文本输入框，用于接收用户输入的水温值（通过 state 状态属性来绑定）。然后，为文本输入框定义"onChange"事件监控方法，实现对水温值的监控。最后，通过【代码 10-1】实现的判断水温级别的函数组件（WaterTempLevel）来渲染水温级别判断结果，具体代码如下：

【代码 10-2】（详见源代码目录 ch10-react-watch-water-temp-level.html 文件）

```
01  <!DOCTYPE html>
02  <html>
03  <head>
04      <meta charset="UTF-8"/>
05      <title>React Event - Water Temperature</title>
06      <script src="https://unpkg.com/react@16/umd/react.development.js"></script>
07      <script src="https://unpkg.com/react-dom@16/umd/react-dom.development.js"></script>
08      <!-- Don't use this in production: -->
09      <script src="https://unpkg.com/babel-standalone@6.15.0/babel.min.js"></script>
10  </head>
11  <body>
12  <!-- 添加文档主体内容 -->
13  <div id='id-div-react'></div>
14  <script type="text/babel">
15      // TODO: define water temperature func Component
16      var divReact = document.getElementById('id-div-react');
17      // TODO: define func Component
18      function WaterTempLevel(props) {
19          if (props.wlevel <= 0) {
20              return <p>This is ice water.</p>;
21          } else if((props.wlevel > 0) && (props.wlevel <= 20)) {
22              return <p>This is cold water.</p>;
23          } else if((props.wlevel > 20) && (props.wlevel <= 38)) {
```

```
24             return <p>This is warm water.</p>;
25         } else if((props.wlevel > 38) && (props.wlevel < 100)) {
26             return <p>This is hot water.</p>;
27         } else if(props.wlevel >= 100) {
28             return <p>This is boiling water.</p>;
29         } else {
30             return <p>This is ... water.</p>;
31         }
32     }
33     // TODO: define ES6 Class Component
34     class WaterTempCelsius extends React.Component {
35         // TODO: constructor
36         // state - temperature
37         constructor(props) {
38             super(props);
39             this.state = {
40                 temperature: ''
41             };
42             this.handleTempChange = this.handleTempChange.bind(this);
43         }
44         // TODO: event
45         handleTempChange(e) {
46             this.setState({
47                 temperature: e.target.value
48             });
49         }
50         // TODO: render
51         render() {
52             const temperature = this.state.temperature;
53             return (
54                 <fieldset>
55                     <legend>Enter temperature in Celsius:</legend>
56                     <input
57                         value={temperature}
58                         onChange={this.handleTempChange} />
59                     <WaterTempLevel
60                         wlevel={parseFloat(temperature)} />
61                 </fieldset>
62             );
63         }
64     }
65     // TODO: React JSX
66     const reactSpan = (
67         <span>
68             <h3>React State Enhance - Water Temperature</h3>
69             <WaterTempCelsius />
70         </span>
```

```
71          );
72          // TODO: React render
73          ReactDOM.render(reactSpan, divReact);
74      </script>
75  </body>
76  </html>
```

关于【代码10-2】的说明：

- 本段代码的核心部分就是第 34~64 行代码实现的类组件（WaterTempCelsius），里面的知识点在前面的章节中均有介绍，这里就不再详细解释了。

测试网页的效果如图 10.1、图 10.2 和图 10.3 所示。如图中的箭头所示，类组件（WaterTempCelsius）会根据用户输入的水温值判断出水温级别。

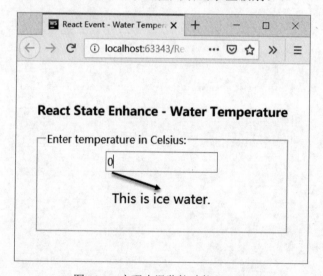

图 10.1 实现水温监控功能（一）

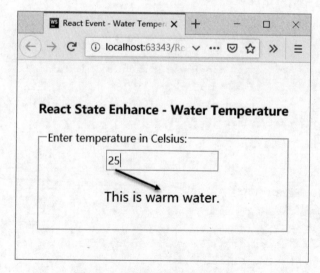

图 10.2 实现水温监控功能（二）

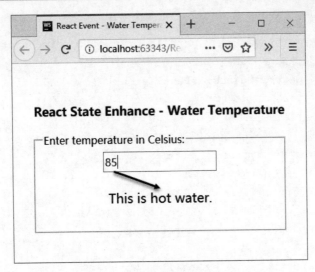

图 10.3 实现水温监控功能（三）

10.3 加入第二个水温输入框

现在回到最初的设计目标，我们要实现一个在不同温度计量标准下同步的功能。此时，我们会想到在【代码 10-2】的基础上再加入第二个水温输入框。第一个水温输入框采用的是摄氏温度计量标准，第二个就选用华氏温度计量标准进行测试，具体代码如下：

【代码 10-3】（详见源代码目录 ch10-react-watch-water-temp-level-2.html 文件）

```
01  <!DOCTYPE html>
02  <html>
03  <head>
04      <meta charset="UTF-8"/>
05      <title>React Event - Water Temperature</title>
06      <script src="https://unpkg.com/react@16/umd/react.development.js"></script>
07      <script src="https://unpkg.com/react-dom@16/umd/react-dom.development.js"></script>
08      <!-- Don't use this in production: -->
09      <script src="https://unpkg.com/babel-standalone@6.15.0/babel.min.js"></script>
10  </head>
11  <body>
12  <!-- 添加文档主体内容 -->
13  <div id='id-div-react'></div>
14  <script type="text/babel">
```

```javascript
15      // TODO: define water temperature func Component
16      var divReact = document.getElementById('id-div-react');
17      // TODO: define func Component
18      function WaterTempCelsiusLevel(props) {
19          if (props.wlevel <= 0) {
20              return <p>This is ice water.</p>;
21          } else if((props.wlevel > 0) && (props.wlevel <= 20)) {
22              return <p>This is cold water.</p>;
23          } else if((props.wlevel > 20) && (props.wlevel <= 38)) {
24              return <p>This is warm water.</p>;
25          } else if((props.wlevel > 38) && (props.wlevel < 100)) {
26              return <p>This is hot water.</p>;
27          } else if(props.wlevel >= 100) {
28              return <p>This is boiling water.</p>;
29          } else {
30              return <p>This is ... water.</p>;
31          }
32      }
33      function WaterTempFahrenheitLevel(props) {
34          if (props.wlevel <= 32) {
35              return <p>This is ice water.</p>;
36          } else if((props.wlevel > 32) && (props.wlevel <= 68)) {
37              return <p>This is cold water.</p>;
38          } else if((props.wlevel > 68) && (props.wlevel <= 100)) {
39              return <p>This is warm water.</p>;
40          } else if((props.wlevel > 100) && (props.wlevel < 212)) {
41              return <p>This is hot water.</p>;
42          } else if(props.wlevel >= 212) {
43              return <p>This is boiling water.</p>;
44          } else {
45              return <p>This is ... water.</p>;
46          }
47      }
48      // TODO: define ES6 Class Component
49      class WaterTemperature extends React.Component {
50          // TODO: constructor
51          // state - temperature
52          constructor(props) {
53              super(props);
54              this.state = {
55                  temperatureCelsius: '',
56                  temperatureFahrenheit: ''
57              };
```

```
58        this.handleTempCelsiusChange =
this.handleTempCelsiusChange.bind(this);
59        this.handleTempFahrenheitChange =
this.handleTempFahrenheitChange.bind(this);
60      }
61      // TODO: handle event
62      handleTempCelsiusChange(e) {
63          this.setState({
64              temperatureCelsius: e.target.value
65          });
66      }
67      handleTempFahrenheitChange(e) {
68          this.setState({
69              temperatureFahrenheit: e.target.value
70          });
71      }
72      // TODO: render
73      render() {
74          const temperatureCelsius = this.state.temperatureCelsius;
75          const temperatureFahrenheit = this.state.temperatureFahrenheit;
76          return (
77              <div>
78                  <span>
79                      <label>Enter temperature in Celsius:</label>
80                      <input
81                          name="c"
82                          value={temperatureCelsius}
83                          onChange={this.handleTempCelsiusChange} />
84                      <WaterTempCelsiusLevel
85                          wlevel={temperatureCelsius} />
86                  </span>
87                  <span>
88                      <label>Enter temperature in Fahrenheit:</label>
89                      <input
90                          name="f"
91                          value={temperatureFahrenheit}
92                          onChange={this.handleTempFahrenheitChange} />
93                      <WaterTempFahrenheitLevel
94                          wlevel={temperatureFahrenheit} />
95                  </span>
96              </div>
97          );
98      }
```

```
99      }
100     // TODO: React JSX
101     const reactSpan = (
102         <span>
103             <h3>React State Enhance - Water Temperature</h3>
104             <WaterTemperature />
105         </span>
106     );
107     // TODO: React render
108     ReactDOM.render(reactSpan, divReact);
109 </script>
110 </body>
111 </html>
```

关于【代码 10-3】的说明：

- 为了设计加入第二个水温输入框，第 80～83 行代码和第 89～92 行代码分别定义了一个文本输入框（摄氏温度和华氏温度）。常规做法是为这两个文本输入框定义各自的 name 属性（c 和 f），value 属性（temperatureCelsius 和 temperatureFahrenheit），以及事件处理方法（handleTempCelsiusChange 和 handleTempFahrenheitChange）。
- 而对于判断水温级别的函数组件而言，由于每个文本输入框的 value 属性是通过各自的状态属性（temperatureCelsius 和 temperatureFahrenheit）进行同步的，故而第 18～32 行代码和第 33～47 行代码分别定义了各自的判断水温级别的函数组件（WaterTempCelsiusLevel 和 WaterTempFahrenheitLevel）。

测试网页的效果如图 10.4 和图 10.5 所示。如图中的箭头所示，这两个水温输入框的数值是无法同步的，只支持各自单独方式的输入。

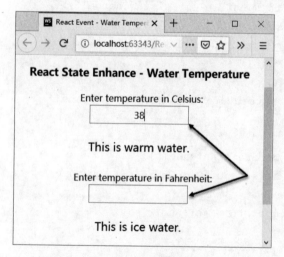

图 10.4　加入第二个温度输入框（一）

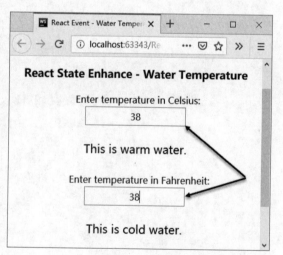

图 10.5　加入第二个温度输入框（二）

10.4 同步二个水温输入框

我们还是要回到最初的设计目标，实现一个在不同温度计量标准下同步的功能。注意到，【代码 10-3】中的两个水温输入框无法同步的主要原因是使用了各自的状态属性（temperatureCelsius 和 temperatureFahrenheit）。于是，可以将两个水温输入框的 value 属性定义为同一个状态属性（temperature），具体代码如下：

【代码 10-4】（详见源代码目录 ch10-react-watch-water-temp-level-3.html 文件）

```
01  <!DOCTYPE html>
02  <html>
03  <head>
04      <meta charset="UTF-8"/>
05      <title>React Event - Water Temperature</title>
06      <script src="https://unpkg.com/react@16/umd/react.development.js"></script>
07      <script src="https://unpkg.com/react-dom@16/umd/react-dom.development.js"></script>
08      <!-- Don't use this in production: -->
09      <script src="https://unpkg.com/babel-standalone@6.15.0/babel.min.js"></script>
10  </head>
11  <body>
12  <!-- 添加文档主体内容 -->
13  <div id='id-div-react'></div>
14  <script type="text/babel">
15      // TODO: define water temperature func Component
16      var divReact = document.getElementById('id-div-react');
17      // TODO: define func Component
18      function WaterTempCelsiusLevel(props) {
19          if (props.wlevel <= 0) {
20              return <p>This is ice water.</p>;
21          } else if((props.wlevel > 0) && (props.wlevel <= 20)) {
22              return <p>This is cold water.</p>;
23          } else if((props.wlevel > 20) && (props.wlevel <= 38)) {
24              return <p>This is warm water.</p>;
25          } else if((props.wlevel > 38) && (props.wlevel < 100)) {
26              return <p>This is hot water.</p>;
27          } else if(props.wlevel >= 100) {
28              return <p>This is boiling water.</p>;
29          } else {
30              return <p>This is ... water.</p>;
```

```
31        }
32     }
33     function WaterTempFahrenheitLevel(props) {
34        if (props.wlevel <= 32) {
35           return <p>This is ice water.</p>;
36        } else if((props.wlevel > 32) && (props.wlevel <= 68)) {
37           return <p>This is cold water.</p>;
38        } else if((props.wlevel > 68) && (props.wlevel <= 100)) {
39           return <p>This is warm water.</p>;
40        } else if((props.wlevel > 100) && (props.wlevel < 212)) {
41           return <p>This is hot water.</p>;
42        } else if(props.wlevel >= 212) {
43           return <p>This is boiling water.</p>;
44        } else {
45           return <p>This is ... water.</p>;
46        }
47     }
48     // TODO: define ES6 Class Component
49     class WaterTemperature extends React.Component {
50        // TODO: constructor
51        // state - temperature
52        constructor(props) {
53           super(props);
54           this.state = {
55              temperature: ''
56           };
57           this.handleTempChange = this.handleTempChange.bind(this);
58        }
59        // TODO: handle event
60        handleTempChange(e) {
61           this.setState({
62              temperature: e.target.value
63           });
64        }
65        // TODO: render
66        render() {
67           const temperature = this.state.temperature;
68           return (
69              <div>
70                 <span>
71                    <label>Enter temperature in Celsius:</label>
72                    <input
73                       name="c"
```

```
74                    value={temperature}
75                    onChange={this.handleTempChange} />
76                <WaterTempCelsiusLevel
77                    wlevel={temperature} />
78            </span>
79            <span>
80                <label>Enter temperature in Fahrenheit:</label>
81                <input
82                    name="f"
83                    value={temperature}
84                    onChange={this.handleTempChange} />
85                <WaterTempFahrenheitLevel
86                    wlevel={temperature} />
87            </span>
88         </div>
89       );
90     }
91   }
92   // TODO: React JSX
93   const reactSpan = (
94     <span>
95        <h3>React State Enhance - Water Temperature</h3>
96        <WaterTemperature />
97     </span>
98   );
99   // TODO: React render
100    ReactDOM.render(reactSpan, divReact);
101 </script>
102 </body>
103 </html>
```

关于【代码 10-4】的说明：

- 第 72～75 行代码和第 81～84 行代码定义的水温输入框（摄氏温度和华氏温度）中的 value 属性、引用了相同的 state 状态属性（temperature），以及相同的事件处理方法（handleTempChange）。
- 第 76～77 行代码和第 85～86 行代码引用的是判断水温级别的函数组件（WaterTempCelsiusLevel 和 WaterTempFahrenheitLevel）中，wlevel 属性也同样引用了 state 状态属性（temperature）。

测试网页的效果如图 10.6 和图 10.7 所示。如图中的箭头所示，这两个水温输入框的数值是可以同步了，但是由于摄氏温度和华氏温度的计量不同，因此华氏温度的数值是不正确的。

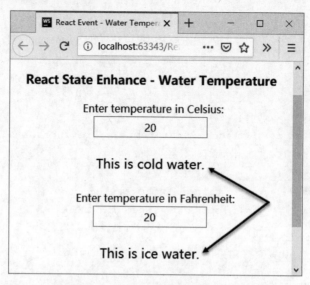

图 10.6　同步二个水温输入框（一）

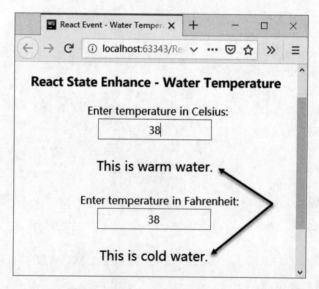

图 10.7　同步二个水温输入框（二）

为了解决这个问题，需要在组件内部进行摄氏温度值和华氏温度值的换算。但是，由于两个水温输入框引用了通过一个 state 状态属性（temperature），所以实现换算的操作会很复杂。因此，需要转换思路，重新架构组件。

10.5　将水温"状态共享"

我们的目标是希望两个水温输入框的数值能够同步（符合摄氏温度和华氏温度的换算规则），当用户在摄氏温度输入框内更新数值时，华氏温度输入框内的数值应同步转换并更新，反之亦然。

首先，我们可以将摄氏温度输入和华氏温度输入定义为两个组件。同时，因为温度需要同步转换，所以需要将温度（temperature）作为 state 状态属性进行共享。请看下面的代码实例：

【代码 10-5】（详见源代码目录 ch10-react-state-enhance-water-temp-1.html 文件）

```
01    <!DOCTYPE html>
02    <html>
03    <head>
04        <meta charset="UTF-8"/>
05        <title>React Event - Water Temperature</title>
06        <script src="https://unpkg.com/react@16/umd/react.development.js"></script>
07        <script src="https://unpkg.com/react-dom@16/umd/react-dom.development.js"></script>
08        <!-- Don't use this in production: -->
09        <script src="https://unpkg.com/babel-standalone@6.15.0/babel.min.js"></script>
10    </head>
11    <body>
12    <!-- 添加文档主体内容 -->
13    <div id='id-div-react'></div>
14    <script type="text/babel">
15        // TODO: define water temperature func Component
16        var divReact = document.getElementById('id-div-react');
17        // TODO: define temperature type
18        const tempType = {
19            c: 'Celsius',
20            f: 'Fahrenheit'
21        };
22        // TODO: define func Component
23        function WaterTempLevel(props) {
24            if(props.type === 'c') {
25                return <WaterTempCelsiusLevel wlevel={props.wlevel} />;
26            } else if(props.type === 'f') {
27                return <WaterTempFahrenheitLevel wlevel={props.wlevel} />;
28            } else {}
29        }
30        function WaterTempCelsiusLevel(props) {
31            if (props.wlevel <= 0) {
32                return <p>This is ice water.</p>;
33            } else if((props.wlevel > 0) && (props.wlevel <= 20)) {
34                return <p>This is cold water.</p>;
35            } else if((props.wlevel > 20) && (props.wlevel <= 38)) {
```

```
36          return <p>This is warm water.</p>;
37      } else if((props.wlevel > 38) && (props.wlevel < 100)) {
38          return <p>This is hot water.</p>;
39      } else if(props.wlevel >= 100) {
40          return <p>This is boiling water.</p>;
41      } else {
42          return <p>This is ... water.</p>;
43      }
44  }
45  function WaterTempFahrenheitLevel(props) {
46      if (props.wlevel <= 32) {
47          return <p>This is ice water.</p>;
48      } else if((props.wlevel > 32) && (props.wlevel <= 68)) {
49          return <p>This is cold water.</p>;
50      } else if((props.wlevel > 68) && (props.wlevel <= 100)) {
51          return <p>This is warm water.</p>;
52      } else if((props.wlevel > 100) && (props.wlevel < 212)) {
53          return <p>This is hot water.</p>;
54      } else if(props.wlevel >= 212) {
55          return <p>This is boiling water.</p>;
56      } else {
57          return <p>This is ... water.</p>;
58      }
59  }
60  // TODO: define ES6 Class Component
61  class WaterTemperature extends React.Component {
62      // TODO: constructor
63      // state - temperature
64      constructor(props) {
65          super(props);
66          this.state = {
67              temperature: ''
68          };
69          this.handleTempChange = this.handleTempChange.bind(this);
70      }
71      // TODO: handle event
72      handleTempChange(e) {
73          let v = e.target.value;
74          this.setState({
75              temperature: v
76          });
77      }
78      // TODO: render
```

```
79      render() {
80          const temperature = this.state.temperature;
81          const type = this.props.type;
82          return (
83              <div>
84                  <span>
85                      <label>Enter temperature in {tempType[type]}:</label>
86                      <input
87                          name={type}
88                          value={temperature}
89                          onChange={this.handleTempChange} />
90                      <WaterTempLevel
91                          wlevel={temperature}
92                          type={type} />
93                  </span>
94              </div>
95          );
96      }
97  }
98  // TODO: define ES6 Class Component
99  class WaterTemperatureApp extends React.Component {
100     // TODO: render
101     render() {
102         return (
103             <div>
104                 <WaterTemperature type="c" />
105                 <WaterTemperature type="f" />
106             </div>
107         );
108     }
109 }
110 // TODO: React JSX
111 const reactSpan = (
112     <span>
113         <h3>React State Enhance - Water Temperature</h3>
114         <WaterTemperatureApp />
115     </span>
116 );
117 // TODO: React render
118 ReactDOM.render(reactSpan, divReact);
119 </script>
120 </body>
121 </html>
```

关于【代码 10-5】的说明：

- 第 61~97 行代码定义了一个水温输入组件（WaterTemperature），支持用户输入水温值，并自动判断温度级别。
- 第 99~109 行代码定义了一个水温输入组件容器（WaterTemperatureApp），其中第 104 行代码和第 105 行代码分别通过引入水温输入组件（WaterTemperature），实现了摄氏温度输入和华氏温度输入功能。
- 这样设计的初衷就是让摄氏温度输入和华氏温度输入可以共享一个 state 状态属性（temperature）。

测试网页的效果如图 10.8 所示。虽然摄氏温度输入和华氏温度输入共享了 state 状态属性，但此时二者还是无法同步。

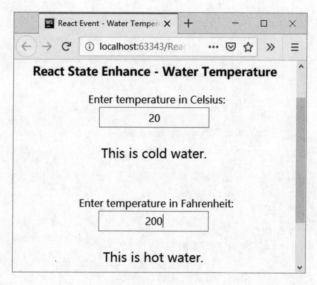

图 10.8　将水温"状态"共享

10.6 将水温"状态提升"

上一节实现了水温状态属性（temperature）的共享，但还无法将摄氏温度输入和华氏温度输入进行同步。此时，就需要将温度（temperature）作为 state 状态属性提升到共同的父组件中去，这个操作就是所谓的 React 框架"状态提升"了。请看下面的代码实例：

【代码 10-6】（详见源代码目录 ch10-react-state-enhance-water-temp-2.html 文件）

```
01  <!DOCTYPE html>
02  <html>
03  <head>
04      <meta charset="UTF-8"/>
```

```
05      <title>React Event - Water Temperature</title>
06      <script src="https://unpkg.com/react@16/umd/react.development.js"></script>
07      <script src="https://unpkg.com/react-dom@16/umd/react-dom.development.js"></script>
08      <!-- Don't use this in production: -->
09      <script src="https://unpkg.com/babel-standalone@6.15.0/babel.min.js"></script>
10    </head>
11    <body>
12    <!-- 添加文档主体内容 -->
13    <div id='id-div-react'></div>
14    <script type="text/babel">
15        // TODO: define water temperature func Component
16        var divReact = document.getElementById('id-div-react');
17        // TODO: define temperature type
18        const tempType = {
19            c: 'Celsius',
20            f: 'Fahrenheit'
21        };
22        // TODO: define func Component
23        function WaterTempLevel(props) {
24            if(props.type === 'c') {
25                return <WaterTempCelsiusLevel wlevel={props.wlevel} />;
26            } else if(props.type === 'f') {
27                return <WaterTempFahrenheitLevel wlevel={props.wlevel} />;
28            } else {}
29        }
30        function WaterTempCelsiusLevel(props) {
31            if (props.wlevel <= 0) {
32                return <p>This is ice water.</p>;
33            } else if((props.wlevel > 0) && (props.wlevel <= 20)) {
34                return <p>This is cold water.</p>;
35            } else if((props.wlevel > 20) && (props.wlevel <= 38)) {
36                return <p>This is warm water.</p>;
37            } else if((props.wlevel > 38) && (props.wlevel < 100)) {
38                return <p>This is hot water.</p>;
39            } else if(props.wlevel >= 100) {
40                return <p>This is boiling water.</p>;
41            } else {
42                return <p>This is ... water.</p>;
43            }
44        }
```

```
45    function WaterTempFahrenheitLevel(props) {
46        if (props.wlevel <= 32) {
47            return <p>This is ice water.</p>;
48        } else if((props.wlevel > 32) && (props.wlevel <= 68)) {
49            return <p>This is cold water.</p>;
50        } else if((props.wlevel > 68) && (props.wlevel <= 100)) {
51            return <p>This is warm water.</p>;
52        } else if((props.wlevel > 100) && (props.wlevel < 212)) {
53            return <p>This is hot water.</p>;
54        } else if(props.wlevel >= 212) {
55            return <p>This is boiling water.</p>;
56        } else {
57            return <p>This is ... water.</p>;
58        }
59    }
60    // TODO: define ES6 Class Component
61    class WaterTemperature extends React.Component {
62        // TODO: constructor
63        // state - temperature
64        constructor(props) {
65            super(props);
66            this.handleTempChange = this.handleTempChange.bind(this);
67        }
68        // TODO: handle event
69        handleTempChange(e) {
70            this.props.onTemperatureChange(e.target.value);
71        }
72        // TODO: render
73        render() {
74            const temperature = this.props.temperature;
75            const type = this.props.type;
76            return (
77                <div>
78                    <span>
79                        <label>Enter temperature in {tempType[type]}:</label>
80                        <input
81                            name={type}
82                            value={temperature}
83                            onChange={this.handleTempChange} />
84                        <WaterTempLevel
85                            wlevel={temperature}
86                            type={type} />
87                    </span>
```

```
 88                </div>
 89            );
 90        }
 91    }
 92    // TODO: define ES6 Class Component
 93    class WaterTemperatureApp extends React.Component {
 94        // TODO: constructor
 95        // state - temperature
 96        constructor(props) {
 97            super(props);
 98            this.state = {
 99                temperature: ''
100            };
101            this.handleCelsiusChange = this.handleCelsiusChange.bind(this);
102            this.handleFahrenheitChange = this.handleFahrenheitChange.bind(this);
103        }
104        // TODO: handle event
105        handleCelsiusChange(temperature) {
106            this.setState({
107                type: 'c',
108                temperature
109            });
110        }
111        handleFahrenheitChange(temperature) {
112            this.setState({
113                type: 'f',
114                temperature
115            });
116        }
117        // TODO: render
118        render() {
119            const temperature = this.state.temperature;
120            return (
121                <div>
122                    <WaterTemperature
123                        type="c"
124                        temperature={temperature}
125                        onTemperatureChange={this.handleCelsiusChange} />
126                    <WaterTemperature
127                        type="f"
128                        temperature={temperature}
```

```
129                        onTemperatureChange={this.handleFahrenheitChange} />
130             </div>
131         );
132     }
133 }
134 // TODO: React JSX
135 const reactSpan = (
136     <span>
137         <h3>React State Enhance - Water Temperature</h3>
138         <WaterTemperatureApp />
139     </span>
140 );
141 // TODO: React render
142 ReactDOM.render(reactSpan, divReact);
143 </script>
144 </body>
145 </html>
```

关于【代码 10-6】的说明：

- 请先看第 93~133 行代码定义的水温输入组件容器（WaterTemperatureApp），相比于【代码 10-5】的定义，这里将水温状态属性（temperature）移进来了，该操作就是"状态提升"。
- 第 122~125 行代码和第 126~129 行代码定义的摄氏温度输入组件和华氏温度输入组件，引用了相同的水温状态属性（temperature），同时还添加了自定义 onChange 事件处理方法（onTemperatureChange）。
- 第 061~091 行代码定义的水温输入组件（WaterTemperature），通过 props 参数获取了水温状态属性（temperature），还有温度类别（type）。
- 关于温度类别(type)，请看第 018~021 行代码定义的温度类别常量对象(tempType)，在水温输入组件（WaterTemperature）通过该对象（tempType）来区分摄氏温度输入组件和华氏温度输入组件。

测试网页的效果如图 10.9 和图 10.10 所示。如图中的箭头所示，摄氏温度输入和华氏温度输入实现了同步。但我们知道摄氏温度和华氏温度是有换算关系的，后面章节会进行讲解。

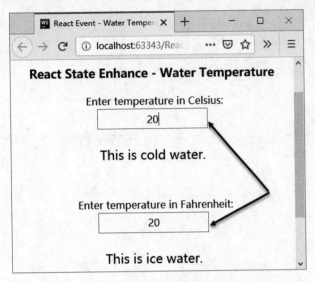

图 10.9　将水温"状态提升"（一）

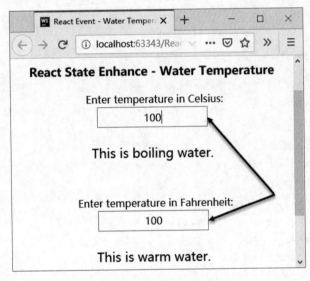

图 10.10　将水温"状态提升"（二）

10.7　实现水温同步换算

本节就是关于"状态提升"的关键内容了，我们将实现摄氏温度输入和华氏温度输入的同步，也包括同步换算功能，请看下面的代码实例：

【代码 10-7】（详见源代码目录 ch10-react-state-enhance-water-temp-3.html 文件）

```
01  <!DOCTYPE html>
02  <html>
```

```html
03  <head>
04      <meta charset="UTF-8"/>
05      <title>React Event - Water Temperature</title>
06      <script src="https://unpkg.com/react@16/umd/react.development.js"></script>
07      <script src="https://unpkg.com/react-dom@16/umd/react-dom.development.js"></script>
08      <!-- Don't use this in production: -->
09      <script src="https://unpkg.com/babel-standalone@6.15.0/babel.min.js"></script>
10  </head>
11  <body>
12  <!-- 添加文档主体内容 -->
13  <div id='id-div-react'></div>
14  <script type="text/babel">
15      // TODO: define water temperature func Component
16      var divReact = document.getElementById('id-div-react');
17      // TODO: define temperature type
18      const tempType = {
19          c: 'Celsius',
20          f: 'Fahrenheit'
21      };
22      // TODO: convert temperature
23      function toCelsius(fahrenheit) {
24          return Math.round((fahrenheit - 32) * 5 / 9);
25      }
26      function toFahrenheit(celsius) {
27          return Math.round((celsius * 9 / 5) + 32);
28      }
29      function toConvert(temperature, convert) {
30          if (Number.isNaN(temperature))
31              return '';
32          return convert(temperature).toString();
33      }
34      // TODO: define func Component
35      function WaterTempLevel(props) {
36          if(props.type === 'c') {
37              return <WaterTempCelsiusLevel wlevel={props.wlevel} />;
38          } else if(props.type === 'f') {
39              return <WaterTempFahrenheitLevel wlevel={props.wlevel} />;
40          } else {}
41      }
42      function WaterTempCelsiusLevel(props) {
```

```
43          if (props.wlevel <= 0) {
44              return <p>This is ice water.</p>;
45          } else if((props.wlevel > 0) && (props.wlevel <= 20)) {
46              return <p>This is cold water.</p>;
47          } else if((props.wlevel > 20) && (props.wlevel <= 38)) {
48              return <p>This is warm water.</p>;
49          } else if((props.wlevel > 38) && (props.wlevel < 100)) {
50              return <p>This is hot water.</p>;
51          } else if(props.wlevel >= 100) {
52              return <p>This is boiling water.</p>;
53          } else {
54              return <p>This is ... water.</p>;
55          }
56      }
57      function WaterTempFahrenheitLevel(props) {
58          if (props.wlevel <= 32) {
59              return <p>This is ice water.</p>;
60          } else if((props.wlevel > 32) && (props.wlevel <= 68)) {
61              return <p>This is cold water.</p>;
62          } else if((props.wlevel > 68) && (props.wlevel <= 100)) {
63              return <p>This is warm water.</p>;
64          } else if((props.wlevel > 100) && (props.wlevel < 212)) {
65              return <p>This is hot water.</p>;
66          } else if(props.wlevel >= 212) {
67              return <p>This is boiling water.</p>;
68          } else {
69              return <p>This is ... water.</p>;
70          }
71      }
72      // TODO: define ES6 Class Component
73      class WaterTemperature extends React.Component {
74          // TODO: constructor
75          constructor(props) {
76              super(props);
77              this.handleTempChange = this.handleTempChange.bind(this);
78          }
79          // TODO: handle event
80          handleTempChange(e) {
81              this.props.onTemperatureChange(e.target.value);
82          }
83          // TODO: render
84          render() {
85              const temperature = this.props.temperature;
```

```
86            const type = this.props.type;
87            return (
88              <div>
89                <span>
90                  <label>Enter temperature in {tempType[type]}:</label>
91                  <input
92                    value={temperature}
93                    onChange={this.handleTempChange} />
94                  <WaterTempLevel
95                    wlevel={temperature}
96                    type={type} />
97                </span>
98              </div>
99            );
100         }
101       }
102       // TODO: define ES6 Class Component
103       class WaterTemperatureApp extends React.Component {
104         // TODO: constructor
105         // state - temperature, type
106         constructor(props) {
107           super(props);
108           this.state = {
109             temperature: '0',
110             type: 'c'
111           };
112           this.handleCelsiusChange = this.handleCelsiusChange.bind(this);
113           this.handleFahrenheitChange = this.handleFahrenheitChange.bind(this);
114         }
115         // TODO: handle event
116         handleCelsiusChange(temperature) {
117           this.setState({type: 'c', temperature});
118         }
119         handleFahrenheitChange(temperature) {
120           this.setState({type: 'f', temperature});
121         }
122         // TODO: render
123         render() {
124           const type = this.state.type;
125           const temperature = this.state.temperature;
126     const celsius = type === 'f' ? toConvert(temperature, toCelsius) :
```

```
temperature;
    127         const fahrenheit = type === 'c' ? toConvert(temperature,
toFahrenheit) : temperature;
    128         return (
    129             <div>
    130                 <WaterTemperature
    131                     type="c"
    132                     temperature={celsius}
    133                     onTemperatureChange={this.handleCelsiusChange} />
    134                 <WaterTemperature
    135                     type="f"
    136                     temperature={fahrenheit}
    137                     onTemperatureChange={this.handleFahrenheitChange} />
    138             </div>
    139         );
    140     }
    141 }
    142 // TODO: React JSX
    143 const reactSpan = (
    144     <span>
    145         <h3>React State Enhance - Water Temperature</h3>
    146         <WaterTemperatureApp />
    147     </span>
    148 );
    149 // TODO: React render
    150 ReactDOM.render(reactSpan, divReact);
    151 </script>
    152 </body>
    153 </html>
```

关于【代码 10-7】的说明：

- 请先看第 103～141 行代码定义的水温输入组件容器（WaterTemperatureApp），相比于【代码 10-6】做了一些修改，具体内容如下：
  - 第 126 行代码通过三元条件运算表达式判断温度输入组件的类型（type），并调用 toConvert() 方法将华氏温度数值转换为摄氏温度数值（celsius）。
  - 第 127 行代码同样通过三元条件运算表达式判断温度输入组件的类型（type），不同的是调用 toConvert() 方法将摄氏温度数值转换为华氏温度数值（fahrenheit）。
  - 第 132 行代码和第 136 行代码分别将温度数值（celsius 和 fahrenheit）传递给温度属性（temperature），并通过 props 参数传递给温度组件（WaterTemperature）。
- 关于温度换算的 toConvert() 方法，参看第 23～33 行代码定义的 3 个函数方法（toConvert()、toCelsius() 和 toFahrenheit()）。

测试网页的效果如图 10.11 至图 10.13 所示。如图中的箭头所示，摄氏温度输入和华氏温度输入实现了同步，包括同步换算功能。

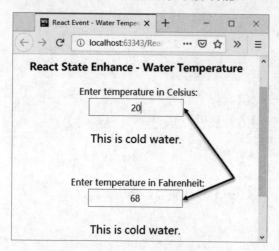

图 10.11　实现水温同步换算（一）

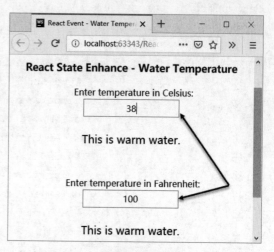

图 10.12　实现水温同步换算（二）

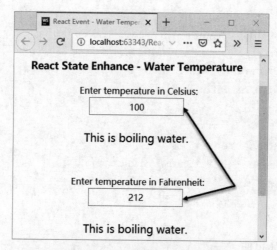

图 10.13　实现水温同步换算（三）

第 11 章

◂React扩展▸

本章主要介绍关于 React 扩展方面的知识，涉及的内容包括 Node、Babel、Webpack、browserify、React Router、单页面以及 Redux 等。这些知识点与 React 环境相关，与 React 扩展相关，都是与 React 设计开发息息相关的内容。

11.1 Node.js 与 React

提起 Node.js 的鼎鼎大名，相信是无人不知、无人不晓。Node.js 凭借其前后端通吃的强大特性，在 Web 开发领域占据着举足轻重的地位。此外，还有基于 Node.js 的 npm 包管理器，也是深受广大 Web 开发人员喜爱的安装、共享、分发和管理代码的工具。

React 作为当下最流行的前端框架，自然也支持通过 npm 包管理工具进行安装，具体的命令如下。

【代码 11-1】

```
// TODO:本地安装
npm install react react-dom --save-dev      // local install react & react-dom
// TODO:全局安装
npm install -g react react-dom              // local install react & react-dom
```

另外，通过使用 npm 包管理工具可以进行项目初始化操作，具体的命令如下：

【代码 11-2】

```
// TODO: npm init
npm init    // initial node project
```

初始化操作后，项目目录下会自生成一个 package.json 文件。该文件十分重要，是配置各种依赖模块的关键所在。

11.2 Babel 与 React

在本书的开篇介绍过关于 Babel 的概念，这里我们再深入介绍 Babel 与 React 二者之间的关系。

严格来讲，Babel 与 React 并不是相互包含的关系，甚至说它们没有直接"依赖关系"也没有问题。Babel 本质就是一个 JavaScript 编译器，主要用于将 ES6+（2015+）版本的代码转换为兼容 ES5 版本的 JavaScript 语法，以便保证最新的 ES6 语法在旧版本浏览器或其他环境中的运行兼容性。

鉴于以上的描述，可以说 Babel 与 React 之间又有着一定的关系。因为，React 框架是基于 ES6 语法构建的，使用了其独有的 JSX 语法。这就需要 Babel 对 React 代码中的 ES6 与 JSX 语法进行转换，以满足旧版本浏览器的兼容性。

总之，Babel 最初并不是专为 React 所设计的，但由于其主要功能是将 ES6+语法转换为 ES5 语法，因此在 React 代码中就可以放心地使用最新的 ES6+语法功能了。至于对 JSX 语法的转化功能，也就轻易实现了。

对于 Babel 功能的安装、配置与使用，会在后面的内容中详细介绍。这里，我们先看一下 Babel 对于 ES6 语法是如何实现转换的（基于最新的 Babel 7+版本），具体如下：

【代码 11-3】（详见源代码目录 ch11-react-babel-es6\es6.js 文件）

```
// TODO: Babel 输入 --- ES6 箭头函数
[1, 2, 3].map((n) => n * n);
```

下面，借助 Babel 官方网站（https://babeljs.io/）自带的转换工具将【代码 11-3】进行转换，具体结果如下。

【代码 11-4】（详见源代码目录 ch11-react-babel-es6\babel-es6.js 文件）

```
// TODO: Babel 输出 --- ES6 箭头函数
"use strict";
[1, 2, 3].map(function (n) {
    return n * n;
});
```

【代码 11-4】是将【代码 11-3】中的 ES6 箭头函数语法转换成 ES5 普通函数语法（默认严格模式："use strict"），二者功能是完全一致。

接着，我们再看一下 Babel 对于 React 代码是如何实现转换的，具体代码如下：

【代码 11-5】（详见源代码目录 ch11-react-babel-es6\react-code.js 文件）

```
// TODO: define ES6 Class Component
class BabelComp extends React.Component {
    render(){
        return <span>React Babel.</span>;
```

```
    }
};
```

下面，借助 Babel 官方网站（https://babeljs.io/）自带的转换工具将【代码 11-5】进行转换，具体结果如下：

【代码 11-6】（详见源代码目录 ch11-react-babel-es6\babel-react-code.js 文件）

```
// TODO: define ES6 Class Component
var BabelComp =
    /*#__PURE__*/
    function (_React$Component) {
        _inherits(BabelComp, _React$Component);
        function BabelComp() {
            _classCallCheck(this, BabelComp);
    return _possibleConstructorReturn(this,
_getPrototypeOf(BabelComp).apply(this, arguments));
        }
        _createClass(BabelComp, [{
            key: "render",
            value: function render() {
                return React.createElement("span", null, "React Babel.");
            }
        }]);
        return BabelComp;
    }(React.Component);
```

【代码 11-6】是将【代码 11-5】中的 React ES6 Class（BabelComp）代码转换成 ES5 代码形式，注意到转换后的组件（BabelComp）是通过函数方式创建的，而类是通过 createClass() 方法创建的。

11.3 Webpack 模块打包器

首先介绍 Webpack 是什么？Webpack 就是一个 JavaScript 应用程序静态模块打包器。JavaScript 应用程序对于 Webpack 而言，全部文件均被视为模块。Webpack 通过 loader 转换文件，通过 plugin 注入钩子等方式，最终输出由多个文件合并而成的模块化项目。

Webpack 有着非常复杂的配置与使用方式，可以实现十分强大的模块化打包功能。下面，我们介绍一下最基本的 Webpack 打包方法（基于最新的 Webpack 4+版本）。

（1）Webpack 是基于 Node 程序所创建的，因此要先安装配置好 Node 开发环境（请参考 Node 相关文档）。如果 Node 程序安装无误的话，npm 包管理器也会正确安装，然后就可以

通过"npm init"命令初始化 npm 的默认配置了,具体命令如下:

```
// TODO: 创建 Webpack 工作目录
mkdir ch11-webpack-demo
// TODO: 进入 Webpack 工作目录
cd ch11-webpack-demo
// TODO: 初始化 npm 默认配置
npm inti -y          // -y 表示全部使用默认配置
```

此时,工作目录下会自动生成一个 npm 配置文件 package.json,具体如下:

【代码 11-7】(详见源代码目录 ch11-webpack-demo\package.json 文件)

```
{
  "name": "ch11-webpack-demo",
  "version": "1.0.0",
  "description": "",
  "main": "index.js",
  "scripts": {
    "test": "echo \"Error: no test specified\" && exit 1"
  },
  "keywords": [],
  "author": "",
  "license": "ISC",
}
```

(2)安装 Webpack 工具,具体命令如下:

```
// TODO: 安装 Webpack
npm install webpack webpack-cli --save-dev      // --save-dev 表示本地安装
```

此时,npm 配置文件 package.json 会添加进去关于 Webpack 和 webpack-cli 的版本信息,具体代码如下:

【代码 11-8】(详见源代码目录 ch11-webpack-demo\package.json 文件)

```
{
  "name": "ch11-webpack-demo",
  "version": "1.0.0",
  "description": "",
  "main": "index.js",
  "scripts": {
    "test": "echo \"Error: no test specified\" && exit 1"
  },
  "keywords": [],
  "author": "",
  "license": "ISC",
```

```json
"devDependencies": {
  "webpack": "^4.35.3",
  "webpack-cli": "^3.3.5"
}
}
```

（3）在工作目录下创建 src 文件夹，编写两个 JavaScript 模块文件（index.js 和 bar.js），具体代码如下：

【代码 11-9】（详见源代码目录 ch11-webpack-demo\src\index.js 文件）

```
import bar from './bar';
// TODO: call bar()
bar();
```

【代码 11-10】（详见源代码目录 ch11-webpack-demo\src\bar.js 文件）

```
export default function bar() {
    // TODO: write your code
}
```

（4）在工作目录下创建 HTML 页面文件（index.html），具体代码如下：

【代码 11-11】（详见源代码目录 ch11-webpack-demo\index.html 文件）

```html
<!DOCTYPE html>
<html lang="en">
<head>
    <meta charset="UTF-8">
    <title>html webpack page</title>
</head>
<body>
<script src="dist/bundle.js"></script>
</body>
</html>
```

（5）在工作目录下创建 Webpack 配置文件（webpack.config.js），具体代码如下：

【代码 11-12】（详见源代码目录 ch11-webpack-demo\webpack.config.js 文件）

```javascript
const path = require('path');
module.exports = {
    entry: './src/index.js',
    output: {
        path: path.resolve(__dirname, 'dist'),
        filename: 'bundle.js'
    }
};
```

其实，webpack.config.js 的配置项有五个，分别是：入口 entry、出口 output、Loader loader、插件 plugins、模式 mode。

最常用的入口 entry 和出口 output，分别制定在哪里寻找项目依赖的资源文件，把资源文件打包后放在哪个目录下面。插件 plugins 用来增加一些特定功能，如代码压缩等。

（6）接下来就可以在工作目录下通过执行 webpack 命令进行打包了。

```
// TODO：执行 Webpack 模块打包命令
webpack
```

webpack 命令执行完成后，通过 Webpack 模块打包器创建的 JavaScript 模块化程序就基本完成了，其目录结构如图 11.1 所示。如图中的箭头所示，项目自动创建了 dist 目录及 bundle.js 模块化打包文件，该文件就是【代码 11-11】中所引用的 JavaScript 脚本文件。

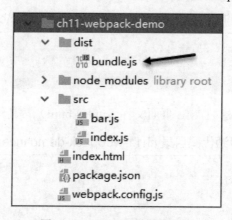

图 11.1　Webpack 模块打包器

【代码 11-13】（详见源代码目录 ch11-webpack-demo\dist\bundle.js 文件）

```
!function(e){var t={};function r(n){if(t[n])return t[n].exports;var
o=t[n]={i:n,l:!1,exports:{}};return
e[n].call(o.exports,o,o.exports,r),o.l=!0,o.exports}r.m=e,r.c=t,r.d=function(e
,t,n){r.o(e,t)||Object.defineProperty(e,t,{enumerable:!0,get:n})},r.r=function
(e){"undefined"!=typeof
Symbol&&Symbol.toStringTag&&Object.defineProperty(e,Symbol.toStringTag,{value:
"Module"}),Object.defineProperty(e,"__esModule",{value:!0})},r.t=function(e,t)
{if(1&t&&(e=r(e)),8&t)return e;if(4&t&&"object"==typeof
e&&e&&e.__esModule)return e;var
n=Object.create(null);if(r.r(n),Object.defineProperty(n,"default",{enumerable:
!0,value:e}),2&t&&"string"!=typeof e)for(var o in e)r.d(n,o,function(t){return
e[t]}.bind(null,o));return n},r.n=function(e){var
t=e&&e.__esModule?function(){return e.default}:function(){return e};return
r.d(t,"a",t),t},r.o=function(e,t){return
Object.prototype.hasOwnProperty.call(e,t)},r.p="",r(r.s=0)}([function(e,t,r){"
use strict";r.r(t)}]);
```

如上面的代码所示，bundle.js 是一个压缩好的 JavaScript 脚本文件。

11.4 Node+Babel+Webpack 搭建 React 环境

前面分别介绍了关于 Node、Babel 和 Webpack 的相关内容，本节将介绍如何通过 Node、Babel 和 Webpack 搭建 Recat 开发环境。

（1）创建一个工作目录（ch11-node-webpack-babel-react），并进入该工作目录下，具体命令如下：

```
// TODO: 创建 Webpack 工作目录
mkdir ch11-node-webpack-babel-react
// TODO: 进入 Webpack 工作目录
cd ch11-webpack-demo
// TODO: 初始化 npm 默认配置
npm inti -y      // -y 表示全部使用默认配置
```

（2）使用 npm 工具初始化默认配置，具体命令如下：

```
// TODO: 初始化 npm 默认配置
npm inti -y      // -y 表示全部使用默认配置
```

此时，工作目录下会自动生成一个 npm 配置文件 package.json，具体代码如下：

【代码 11-14】（详见源代码目录 ch11-node-webpack-babel-react\package.json 文件）

```
{
  "name": "ch11-node-webpack-babel-react",
  "version": "1.0.0",
  "main": "index.js",
  "scripts": {
    "test": "echo \"Error: no test specified\" && exit 1"
  },
  "author": "",
  "license": "ISC",
  "keywords": [],
  "description": ""
}
```

（3）安装配置 Webpack 工具，具体命令如下：

```
// TODO: 安装 Webpack
npm install webpack webpack-cli --save-dev      // --save-dev 表示本地安装
```

此时，npm 配置文件 package.json 会添加进去关于 Webpack 和 webpack-cli 的版本信息，

具体代码如下:

【代码 11-15】(详见源代码目录 ch11-node-webpack-babel-react\package.json 文件)

```json
{
  "name": "ch11-node-webpack-babel-react",
  "version": "1.0.0",
  "main": "index.js",
  "scripts": {
    "test": "echo \"Error: no test specified\" && exit 1"
  },
  "author": "",
  "license": "ISC",
  "keywords": [],
  "description": "",
  "devDependencies": {
    "webpack": "^4.35.3",
    "webpack-cli": "^3.3.5"
  }
}
```

此时,可以在 package.json 配置文件中加入 webpack 命令,具体如下:

```
"build": "webpack --mode production"    // production 表示生产环境
```

备 注
从 Webpack 4+ 版本开始,Webpack 配置文件不再是必需的了。

(4)配置 Babel 环境支持。由于 React 是基于 ES6 编写的,所以要安装相关的支持插件,具体如下:

```
// TODO: Babel core、env、loader and ES6(2015)
npm install @babel/core babel-loader @babel/preset-env --save-dev
npm install babel-preset-es2015 --save-dev
```

安装完成后,package.json 配置文件会增加相应的配置信息。此时,还需要配置一下 Babel,方法就是在工作目录下创建 .babelrc 文件,添加如下的代码:

【代码 11-16】(详见源代码目录 ch11-node-webpack-babel-react\.babelrc 文件)

```json
{
  "presets": ["@babel/preset-env"]
}
```

(5)此时需要简单配置一下 Webpack 的配置文件,具体代码如下:

【代码11-17】（详见源代码目录 ch11-node-webpack-babel-react\webpack.config.js 文件）

```js
module.exports = {
  module: {
    rules: [
      {
        test: /\.(js|jsx)$/,
        exclude: /node_modules/,
        use: {
          loader: "babel-loader"
        }
      }
    ]
  }
};
```

通过上面的配置，Webpack 对于 js 文件会通过 babel-loader 进行管理，具体就是将 ES6 代码转换为旧版本的 ES5 代码。

（6）安装 React 环境，具体如下：

```
// TODO: React React DOM
npm install react react-dom --save-dev
```

安装完成后，package.json 配置文件会增加相应的配置信息。此时，还需要安装配置一下 Babel 对 React 的支持，具体如下：

```
// TODO: React React DOM
npm install @babel/preset-react --save-dev
```

然后，继续配置一下 .babelrc 文件，添加如下的代码：

【代码11-18】（详见源代码目录 ch11-node-webpack-babel-react\.babelrc 文件）

```
{
  "presets": ["@babel/preset-env", "@babel/preset-react"]
}
```

通过上面的配置，Webpack 对于 jsx 文件会通过 babel-loader 进行管理，具体就是将 JSX 代码转换为旧版本的 ES5 代码。

（7）现在就可以编写 React 组件了。先在工作目录下创建 src 文件夹，再在 src 文件夹下再创建一个 js 文件夹，用于放置 React 模块文件（welcome.jsx），具体代码如下：

【代码11-19】（详见源代码目录 ch11-node-webpack-babel-react\src\js\welcome.jsx 文件）

```
01  import React from 'react';
02  import ReactDOM from 'react-dom';
```

```
03    // TODO: define water temperature func Component
04    var divReact = document.getElementById('id-div-react');
05    // TODO: define ES6 Class Component
06    class WelcomeComp extends React.Component {
07        // TODO: constructor
08        constructor(props) {
09            super(props);
10        }
11        // TODO: render
12        render() {
13            return <p>Hello, webpack+Babel+React!</p>;
14        }
15    }
16    // TODO: render
17    ReactDOM.render(<WelcomeComp />, divReact);
```

关于【代码11-19】的说明：

- 这段 React 代码主要定义了一个很简单的页面组件（WelcomeComp），在页面中显示一行文本信息。
- 需要说明的是第 01 行和第 02 行代码，是通过 import 方式引入的本地 "react" 和 "react-dom" 组件。

（8）定义主入口脚本文件（index.js）。

由于 Webpack 4+版本默认的主入口脚本文件为 "./src/index.js"，所以还需要添加该文件，具体代码如下：

【代码11-20】（详见源代码目录 ch11-node-webpack-babel-react\src\index.js 文件）

```
01    import WelcomeComp from "./js/welcome.jsx";
```

关于【代码11-20】的说明：

- 这行代码主要就是通过 import 方式引入 WelcomeComp 模块。

（9）通过 npm 重新生成应用，具体如下：

```
npm run biuld
```

操作成功后，读者会在 dist 目录下看到新生成的 main.js 文件。该文件会自动生成大量复杂的、经过压缩的脚本代码，此处就不给读者展示了。

（10）加入 HTML 页面文件。

对于一个完整的 React 应用，自然还需要 HTML 文件的支持，用于在浏览器中展示 React 模块组件。Webpack 通过安装插件（html-webpack-plugin、html-loader）的方式实现了对 HTML 的支持，具体如下：

```
npm install html-webpack-plugin html-loader --save-dev
```

安装完成后,还需要更新 Webpack 的配置,具体代码如下:

【代码 11-21】(详见源代码目录 ch11-node-webpack-babel-react\webpack.config.js 文件)

```js
const HtmlWebPackPlugin = require("html-webpack-plugin");
module.exports = {
    module: {
        rules: [
            {
                test: /\.(js|jsx)$/,
                exclude: /node_modules/,
                use: {
                    loader: "babel-loader"
                }
            },
            {
                test: /\.html$/,
                use: [
                    {
                        loader: "html-loader"
                    }
                ]
            }
        ]
    },
    plugins: [
        new HtmlWebPackPlugin({
            template: "./src/index.html",
            filename: "./index.html"
        })
    ]
};
```

经过上面的配置,Webpack 就可以支持 HTML 文件的打包了。下面,在 src 目录下新建一个 index.html 页面文件,具体代码如下:

【代码 11-22】(详见源代码目录 ch11-node-webpack-babel-react\src\index.html 文件)

```html
01  <!DOCTYPE html>
02  <html>
03  <head>
04      <meta charset="UTF-8"/>
05      <title>webpack+Babel+React</title>
06  </head>
```

```
07    <body>
08        <!-- 添加文档主体内容 -->
09        <div id='id-div-react'></div>
10    </body>
11  </html>
```

关于【代码 11-22】的说明：

- 第 09 行代码通过<div>定义了一个层元素，并定义了 id 属性。这个 id 属性很关键，React 组件渲染时会用到。
- 请读者再回去看一下【代码 11-22】中的第 04 行代码，它通过该 id 属性值获取的<div>元素，然后在第 17 行代码中完成的渲染操作。

（11）通过 Webpack 重新生成应用，具体如下：

```
npm run biuld
```

操作成功后，读者会在 dist 目录下看到新生成的 index.html 页面文件，具体代码如下：

【代码 11-23】（详见源代码目录 ch11-node-webpack-babel-react\dist\index.html 文件）

```
01  <!DOCTYPE html>
02  <html>
03  <head>
04      <meta charset="UTF-8"/>
05      <title>webpack+Babel+React</title>
06  </head>
07  <body>
08      <!-- 添加文档主体内容 -->
09      <div id='id-div-react'></div>
10      <script type="text/javascript" src="main.js"></script>
11  </body>
12  </html>
```

关于【代码 11-23】的说明：

- 读者会注意到，dist 目录下的 index.html 文件与原始 src 目录下的 index.html 文件基本一致。唯一的区别就是第 10 行代码通过<srcipt>标签自动引入了 main.js 脚本文件。
- 在使用 Webpack 4+版本时是无需在 HTML 页面文件中包含脚本文件的，Webpack 打包器会将 bundle（模块）将自动插入到页面中。

现在，就可以使用 Firefox 浏览器运行测试 dist 目录下的 index.html 网页了，具体效果如图 11.2 所示。

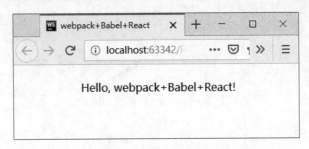

图 11.2 Webpack+Babel+React 环境搭建

如图 11.1 中的箭头所示，在页面中单击按钮后，浏览器控制台中输出了第 19 行代码定义的日志信息。

（12）添加 Web 服务支持。

每次修改应用代码后，都需要重新 biuld 一下的操作，很不友好。好在 Webpack 支持 Web 服务的功能，通过安装 webpack-dev-server 插件就可以实现，具体如下：

```
npm install webpack-dev-server --save-dev
```

然后，在配置文件 package.json 文件中进行相应的修改，具体如下：

```
"start": "webpack-dev-server --open --mode development"
```

配置完成后，通过下面的命令就可以启动 Web 服务了。

```
npm start
```

此时，HTML 页面会通过 HTTP 服务器（默认端口：8080）启动，命令行终端的效果如图 11.3 所示。

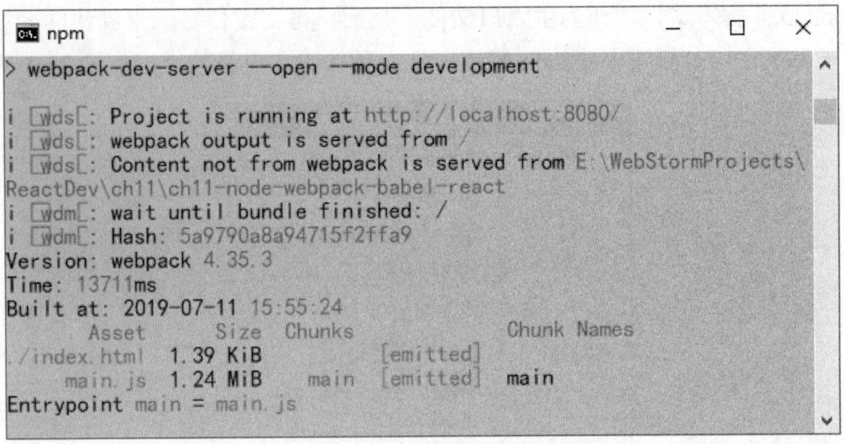

图 11.3 Webpack+Babel+React 应用（命令行）

页面效果如图 11.4 所示。

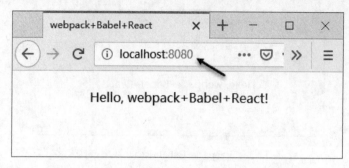

图 11.4　Webpack+Babel+React 应用（Web 方式）

11.5　Browserify 模块打包器

其实还有一款与 Webpack 类似的模块打包器，就是 Browserify。不过，Browserify 自身仅仅自持 Node 模块，对 CSS 样式文件和其他资源文件是不支持的。但这并不影响设计人员对它的喜爱，主要是因为 Browserify 的优雅与纯粹。

下面，我们就简单介绍关于 Browserify 模块打包器的使用方法。

（1）安装 Browserify，具体命令如下：

```
// TODO: 安装 browserify
npm install -g browserify
```

（2）新建一个工作目录（ch11-browserify-demo）用于存放项目文件。首先，在工作目录下新建一个 main.js 脚本文件，作为主入口文件。然后，再在工作目录下创建两个二级子目录和两个三级子目录，同时新建相应的 js 模块文件。具体工作目录的结构如图 11.5 所示。

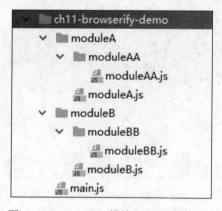

图 11.5　Browserify 模块打包器工作目录

图 11.5 中，包括 main.js 主入口文件在内，一共有 5 个脚本文件。下面，我们分别给出代码：

【代码 11-24】（详见源代码目录 ch11-browserify-demo\main.js 文件）

```
01  /**
02   * entry - main.js
03   */
04  var mA = require('./moduleA/moduleA.js');
05  var mB = require('./moduleB/moduleB.js');
06  console.log(mA + " <-> " + mB);
```

【代码 11-25】（详见源代码目录 ch11-browserify-demo\moduleA\moduleA.js 文件）

```
01  /**
02   * moduleA - moduleA.js
03   */
04  var mAA = require('./moduleAA/moduleAA.js');
05  module.exports = "moduleA" + " <- " + mAA;
```

【代码 11-26】（详见源代码目录 ch11-browserify-demo\moduleA\moduleAA\moduleAA.js 文件）

```
01  /**
02   * moduleA\moduleAA - moduleAA.js
03   */
04  module.exports = "moduleAA";
```

【代码 11-27】（详见源代码目录 ch11-browserify-demo\moduleB\moduleB.js 文件）

```
01  /**
02   * moduleB - moduleB.js
03   */
04  var mBB = require('./moduleBB/moduleBB.js');
05  module.exports = "moduleB" + " <- " + mBB;
```

【代码 11-28】（详见源代码目录 ch11-browserify-demo\moduleB\moduleBB\moduleBB.js 文件）

```
01  /**
02   * moduleB\moduleBB - moduleBB.js
03   */
04  module.exports = "moduleBB";
```

（3）此时在工作目录下通过执行 browserify 命令进行打包，具体如下：

```
// TODO: 执行 browserify 模块打包命令
browserify main.js > .\dist\bundle.js
```

browserify 命令执行完成后，工作目录结构如图 11.6 所示。如图中的箭头所示，项目自动创建了 dist 目录及 bundle.js 模块化打包文件。

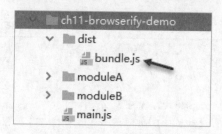

图 11.6 browserify 模块打包器操作结果

具体代码如下：

【代码 11-29】（详见源代码目录 ch11-browserify-demo\dist\bundle.js 文件）

```
    (function(){function r(e,n,t){function o(i,f){if(!n[i]){if(!e[i]){var
c="function"==typeof require&&require;if(!f&&c)return c(i,!0);if(u)return
u(i,!0);var a=new Error("Cannot find module '"+i+"'");throw
a.code="MODULE_NOT_FOUND",a}var
p=n[i]={exports:{}};e[i][0].call(p.exports,function(r){var n=e[i][1][r];return
o(n||r)},p,p.exports,r,e,n,t)}return n[i].exports}for(var u="function"==typeof
require&&require,i=0;i<t.length;i++)o(t[i]);return o}return
r})()({1:[function(require,module,exports){
    /**
    * entry - main.js
    */
    var mA = require('./moduleA/moduleA.js');
    var mB = require('./moduleB/moduleB.js');
    console.log(mA + " <-> " + mB);
},{"./moduleA/moduleA.js":2,"./moduleB/moduleB.js":4}],2:[function(require,
module,exports){
    /**
    * moduleA - moduleA.js
    */
    var mAA = require('./moduleAA/moduleAA.js');
    module.exports = "moduleA" + " <- " + mAA;
},{"./moduleAA/moduleAA.js":3}],3:[function(require,module,exports){
/**
    * moduleA\moduleAA - moduleAA.js
    */
    module.exports = "moduleAA";
},{}],4:[function(require,module,exports){
    var mBB = require('./moduleBB/moduleBB.js');
    module.exports = "moduleB" + " <- " + mBB;
},{"./moduleBB/moduleBB.js":5}],5:[function(require,module,exports){
    module.exports = "moduleBB";
},{}]},{},[1]);
```

下面使用 Node 程序运行测试打包生成的脚本文件，如图 11.7 所示。从输出的结果可以看到，通过 browserify 打包后生成的 bundle.js 文件，成功将 main.js 等 5 个脚本文件封装在一起了。

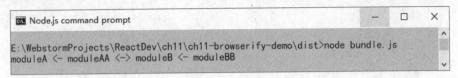

图 11.7　测试打包文件

11.6　React Router 与单页面应用

React Router 就是 React 路由，是 React 官方给出的、完整的 React 框架的路由解决方案。React Router（路由）通过很简单的 API 实现了很强大的功能，通过路由来保持页面 UI 与 URL 地址的同步。

这个 React Router（路由）实现的 UI 与 URL 同步，恰恰是当下流行的单页面应用的基础。所谓"单页面"应用，简单来说就是将视图和数据设计在一个页面中显示，且不同视图的切换也在这个唯一的页面中完成。此时，用户是完全看不到页面发生跳转的。

React Router（路由）是一个基于 React 框架之上的开发出来的库，因此使用时需要通过 npm 包管理器来安装开发包（react-router-dom），安装成功后就可以在应用中添加视图和数据流了。

下面，为了让读者对 React Router 有一些的感性认识，我们通过一个简单的导航视图代码应用进行介绍。

（1）通过 React 脚手架（create-react-app）创建一个应用目录（ch11-react-router-demo），具体方法如下：

```
// TODO: create-react-app
create-react-app ch11-react-router-demo
```

（2）进入该应用目录并安装 React Router 开发包（react-router-dom），方法如下：

```
// TODO: react-router-dom
npm install react-router-dom
```

（3）修改应用目录中 App.js 脚本文件，加入 React Router 代码，具体代码如下：

【代码 11-30】（详见源代码目录 ch05-react-event-onClick.html 文件）

```
01  import React from 'react';
02  import {BrowserRouter as Router, Route, Link} from "react-router-dom";
03  /**
04   * function - Route
```

```
05   */
06  function Index() {
07      return <p>This is home page.</p>;
08  }
09  function About() {
10      return <p>This is about page.</p>;
11  }
12  function Users() {
13      return <p>This is users page.</p>;
14  }
15  /**
16   * function - Router App
17   * @returns {boolean}
18   * @constructor
19   */
20  function App() {
21      return (
22          <Router>
23              <div>
24                  <nav>
25                      <ul>
26                          <li><Link to="/">Home</Link></li>
27                          <li><Link to="/about/">About</Link></li>
28                          <li><Link to="/users/">Users</Link></li>
29                      </ul>
30                  </nav>
31                  <Route path="/" exact component={Index} />
32                  <Route path="/about/" component={About} />
33                  <Route path="/users/" component={Users} />
34              </div>
35          </Router>
36      );
37  }
38  // TODO: export app
39  export default App;
```

关于【代码 11-30】的说明：

- 核心部分就是第 22～35 行代码定义的 Router（路由）功能，具体内容如下：
  - 第 22～35 行代码通过<Router>标签定义了一组路由。
  - 第 25～29 行代码通过<ul><li>标签定义了一组列表，通过<Link>标签内的 "to" 属性指向 URL 路径。
  - 第 31～33 行代码通过<Route>标签定义了三个路由，其路径 "path" 属性对应

<Link>标签内的"to"属性，组件"component"对应第 06~14 行代码定义的三个函数组件。

- 第 06~14 行代码定义了三个函数组件（Index、About 和 Users），分别用于渲染三个导航标签下所对应的视图。

（4）在命令行中输入"npm start"启动 Web 服务器并自动运行该应用，具体效果如图 11.8、图 11.9 和图 11.10 所示。如图中的箭头所示，单击导航链接后，页面导航视图会随之进行切换，说明 React Router（路由）实现了基本的单页面效果。

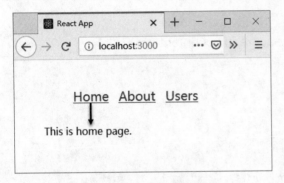

图 11.8　React Router 与单页面应用（一）

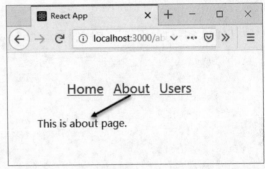

图 11.9　React Router 与单页面应用（二）

图 11.10　React Router 与单页面应用（三）

11.7　Redux 与 React

很多开发者最初接触 Redux 时都是一头雾水，根本搞不清楚 Redux 到底是做什么用的。这主要是因为 Redux 提出了很多全新的概念，比如：store、reducer、dispatch、middleware 等等。

这里，我们不想把官方文档中关于这些概念的释义照搬上来，，我们还是直接以实例应用的方式向读者进行介绍。

不过，有几点关于 Redux 的常识还是要明确一下：

- Redux 其实就是一个 JavaScript 状态容器,实现了可预测的状态管理。
- 虽然读者了解最多的是关于 Redux 和 React 的文章,但 Redux 并不属于 React 框架。其实 Redux 是完全独立的,并不属于任何前端框架,但却可以和大多数前端框架完美结合(恰好目前 Redux 与 React 的结合比较流行)。
- 还有一点提示,学习 Redux 不要拘泥于那些很晦涩的概念(如 store、reducer、dispatch、middleware 等),多做一些应用练习后自然也就理解那些概念了。

下面,为了让读者对 React Router 有一定的感性认识,我们通过一个简单的计数器代码应用进行介绍,具体代码如下:

【代码 11-31】(详见源代码目录 ch11-react-redux-counter.html 文件)

```
01  <!DOCTYPE html>
02  <html>
03  <head>
04      <meta charset="UTF-8" />
05      <title>React Redux Demo</title>
06      <script src="https://unpkg.com/react@16/umd/react.development.js"></script>
07      <script src="https://unpkg.com/react-dom@16/umd/react-dom.development.js"></script>
08      <script src="https://unpkg.com/redux@latest/dist/redux.min.js"></script>
09      <!-- Don't use this in production: -->
10      <script src="https://unpkg.com/babel-standalone@6.15.0/babel.min.js"></script>
11  </head>
12  <body>
13  <!-- 添加文档主体内容 -->
14  <div id='id-div-react'>
15      <p>
16          Clicked: <span id="id-count">0</span> times.<br/><br/>
17          <button id="increment">+1</button>
18          <button id="decrement">-1</button>
19          <button id="incrementIfOdd">+1(当前 times 为奇数时)</button>
20          <button id="incrementAsync">+1(异步操作)</button>
21      </p>
22  </div>
23  <script type="text/babel">
24      // TODO: get div
25      var divReact = document.getElementById('id-div-react');
26      // TODO: define Reducer
27      function reducerCounter(state, action) {
28          if (typeof state === 'undefined') {
29              return 0
30          }
```

```
31        switch (action.type) {
32            case 'INCREMENT':
33                return state + 1
34            case 'DECREMENT':
35                return state - 1
36            default:
37                return state
38        }
39    }
40    // TODO: Create Store
41    var store = Redux.createStore(reducerCounter);
42    var vCount = document.getElementById('id-count');
43    // TODO: render
44    function render() {
45        vCount.innerHTML = store.getState().toString();
46    }
47    render();
48    store.subscribe(render);    // TODO: subscribe
49    // TODO: dispatch
50    document.getElementById('increment').addEventListener('click', function () {
51        store.dispatch({ type: 'INCREMENT' })
52    });
53    document.getElementById('decrement').addEventListener('click', function () {
54        store.dispatch({ type: 'DECREMENT' })
55    });
56    document.getElementById('incrementIfOdd').addEventListener('click', function () {
57        if (store.getState() % 2 !== 0) {
58            store.dispatch({ type: 'INCREMENT' })
59        }
60    });
61    document.getElementById('incrementAsync').addEventListener('click', function () {
62        setTimeout(function () {
63            store.dispatch({ type: 'INCREMENT' })
64        }, 1000)
65    });
66 </script>
67 </body>
68 </html>
```

关于【代码 11-31】的说明：

- 第 08 行代码以 unpkg 的方式（Web 方式）引用 Redux 库文件。这种方式比较简便，具体说明可参考 Redux 官方文档。

- 第16~20行代码定义了一个计数器显示区域和一组（4个）按钮，通过点击不同的按钮来改变计数器的数值。
- 第27~39行代码定义的就是Redux的Reducer对象(通过reducerCounter函数实现)，具体内容如下：
  - 第27行代码的函数名（reducerCounter）中，定义了state参数和action参数。其中，参数state表示状态，参数action表示用户操作的类型。
  - 第28~30行代码通过if条件语句判断state参数是否未定义，如果是，就初始化数值0。
  - 第31~38行代码通过switch条件选择语句判断action.type类型，根据具体类型执行相对应的操作。
- 第41行代码通过Redux的createStore()方法创建了Store对象，其参数为第27~39行代码定义的reducerCounter()函数。
- 第44~46行代码定义了一个渲染函数（render），用于将状态变化反馈到用户视图中，并在第47行代码调用了一次render()函数。
- 第48行代码通过Store对象（store）调用subscribe()方法实现了用户订阅功能，当状态值发生改变时会自动渲染用户视图。
- 第50~65行代码定义了一组（4个）Redux的dispatch操作，对应第17~20行代码定义的4个按钮。当用户点击界面中不同的按钮时，就会触发相对应的dispatch操作。

下面，使用Firefox浏览器运行测试该网页，初始效果如图11.11所示。如图中的箭头所示，页面初始化后浏览器控制台中显示state状态值为0，说明第28~30行代码的if条件语句起到了作用。

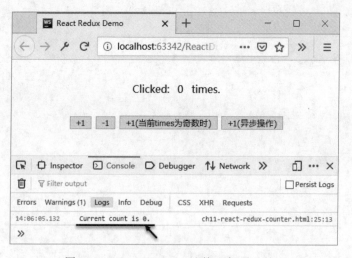

图11.11　React + Redux计数器实现（一）

然后，可以依次点击页面中的几个按钮进行测试，并注意观察页面中的计数器显示以及浏览器控制台中显示的state状态值变化，具体如图11.12、图11.13、图11.14和图11.15所示。

第 11 章 React 扩展

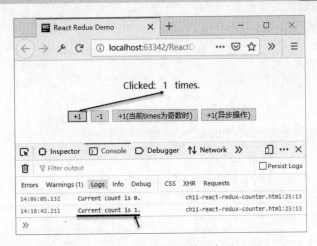

图 11.12　React + Redux 计数器实现（二）

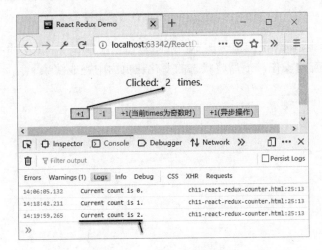

图 11.13　React + Redux 计数器实现（三）

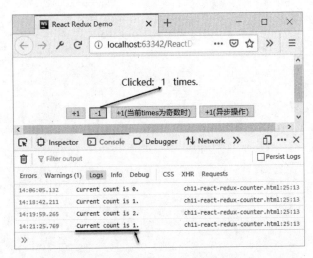

图 11.14　React + Redux 计数器实现（四）

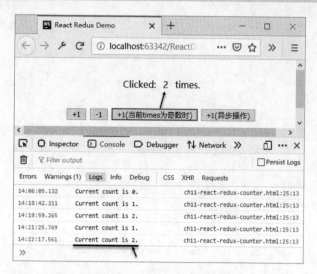

图 11.15　React + Redux 计数器实现（五）

如图 11.12 至图 11.15 中的箭头所示，通过点击按钮触发了 dispatch，然后根据 action 类型来对状态进行相应的操作，最后将状态的变化在页面中完成渲染。

第 12 章

实战1：基于React + Redux 实现计数器应用

本章介绍一个基于 React + Redux 实现的计数器应用，这是一个通过 Redux 结合 React 使用的最基本范例。

12.1 设计思想

Redux 设计之初的目的就是如何更好地维护 Web 页面的状态，而 React 框架恰恰实现了这个 State（状态）的概念，因此 React 与 Redux 是一个很理想的设计组合。

Redux 设计的核心就是 Store 这个概念，它是将 State、Action 和 Reducer 联系在一起的关键。而且特别需要读者注意的一点，Redux 应用中有且仅有一个 Store，也就是说 Store 是唯一的。

Action 是负责把数据从应用传到 Store 的载体，是 Store 数据的唯一来源，具体是通过 Store 对象的 dispatch()方法来实现的。

Reducer 负责将 State（状态）的变化和相应 Action 发送到 Store，然后返回新的 State（状态）。简单来说，就是接收旧的 State（状态）的变化，并返回新的 State（状态）给 Store。

本章介绍的这个计数器应用，就是将计数器的数值作为 React State（状态）来实现，并通过 Redux 来维护这个 State（状态）。这是一个最基础的 Redux 应用，当 Store 发生变化时做了简化处理，此时 React 组件会以手动方式重新渲染。

为了便于读者更好地理解这个示例，我们根据计数器应用的主体架构绘制了一张逻辑图，如图 12.1 所示。

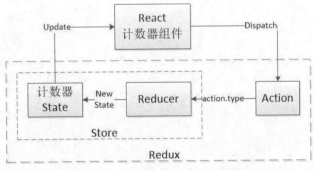

图 12.1　React + Redux 计数器应用逻辑图

如图 12.1 所示，首先要区分的是 Redux 和 React 两大功能模块。概括来讲，Redux 模块负责计数器状态管理，React 负责计数器视图渲染。Redux 模块中的 Store 是核心对象，在管理 Reducer 和 State 的同时，还处理通过 Dispatch 传入的 Action，并最终订阅 React 视图更新。

关于计数器应用的代码结构，如图 12.2 所示。

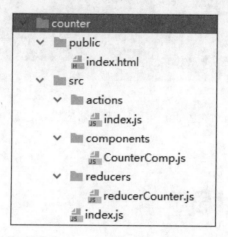

图 12.2　React + Redux 计数器应用结构图

如图 12.2 中所示，文件目录名称（如 actions、components、reducers）基本代表了 js 模块文件的功能，后面我们会逐一介绍这些模块的功能。

12.2　计数器应用页面

可以发现，对于使用 React + Redux 开发的应用来说，HTML 页面在其中的地位会显得不再重要。因为，页面渲染和数据管理工作基本都是通过 React 框架和 Redux 开发包来完成了。

本章实现的计数器应用也是如此，但无论 HTML 页面多么简单，应用中还是需要有一个基本的 HTML 主页的，具体代码如下：

【代码 12-1】（详见源代码 counter 目录下的 public\index.html 文件）

```
01  <!doctype html>
02  <html lang="en">
03    <head>
04      <meta charset="utf-8">
05      <meta name="viewport" content="width=device-width, initial-scale=1">
06      <title>Redux Counter App</title>
07    </head>
08    <body>
09      <div id="root"></div>
10    </body>
```

```
11    </html>
```

关于【代码 12-1】的说明：

- 这段代码很简单，唯一需要说明的就是第 09 行代码通过<div>标签定义的层元素，添加了 id 属性（后面会使用到该属性）。在后面的代码中，React 框架将会把计数器组件渲染到该<div>层内。

12.3 主入口模块

在基于 React + Redux 开发的应用中，需要定义一个主入口模块（js 文件）。在这个主入口模块中，首先要创建一个的 Store 对象（特别强调：Redux 规定 Store 对象是唯一的），然后要渲染视图模型（通过 React 组件实现），最后还要通过 Store 对象实现订阅操作（维护状态数据的更新）。这基本就是主入口模块中需要完成的功能，当然还有更复杂、更高级的功能会在后面进行介绍。

下面是计数器应用主入口模块（js 文件）的内容，具体代码如下：

【代码 12-2】（详见源代码 counter 目录下的 src\index.js 文件）

```
01   import React from 'react';
02   import ReactDOM from 'react-dom';
03   import { createStore } from 'redux';
04   import reducerCounter from './reducers/reducerCounter';
05   import CounterComp from './components/CounterComp';
06   // TODO: get element
07   const rootEle = document.getElementById('root');
08   // TODO: Create Store
09   const store = createStore(reducerCounter);
10   // TODO: render
11   const render = () => ReactDOM.render(
12     <CounterComp
13       value={store.getState()}
14       vstore={store}
15     />,
16     rootEle
17   );
18   // TODO: call render
19   render();
20   // TODO: subscribe
21   store.subscribe(render);
```

关于【代码 12-2】的说明：

- 首先，第 01~05 行代码通过 import 引入了三个组件库（react、react-dom 和 redux）和两个自定义组件，具体说明如下：
  - 第 01 行代码引入了 react 对象（React）。
  - 第 02 行代码引入了 react-dom 对象（ReactDOM）。
  - 第 03 行代码通过 ES6 语法直接引入了 createStore 对象方法，该对象方法是 Redux 的核心所在。
  - 第 04 行代码引入了自定义组件（reducerCounter），这就是所谓的 Reducer 模块组件。
  - 第 05 行代码引入了自定义组件（CounterComp），该组件就是用于页面展示的视图模块组件。
- 第 09 行代码通过调用 createStore()方法获取了 store 对象，注意参数是一个 Reducer 对象（reducerCounter）。
- 第 11~17 行代码通过调用 render()方法完成了页面渲染操作，具体说明如下：
  - 第 12~15 行代码通过自定义组件（CounterComp）实现了计数器。
  - 第 13 行代码定义了一个 value 属性，通过 store 对象的 getState()方法获取了计数器状态值。
  - 第 14 行代码定义了一个 vstore 属性，属性值为 store 对象。
  - 注意，第 13~14 行代码定义的 value 属性和 vstore 属性是作为 Props 参数传递给 CounterComp 组件的，通过该方式将 store 对象传递给子组件。
- 第 21 行代码通过调用 store 对象的 subscribe()方法实现了订阅操作。

说　明
该主入口模块实现了将 Redux 状态管理模块和 React 视图组件模块整合的功能，相当于整个应用的入口。

12.4　视图模块

所谓视图模块，就是用于展示页面功能的组件。在基于 React + Redux 开发的应用中，视图模块主要是通过 React 组件来实现的。

本章的计数器应用中，计数器展示功能是通过 CounterComp 组件完成的。在设计 CounterComp 组件时，首先要考虑如何通过组件特性来维护计数器的核心功能。对于本例的计数器应用，只需要将计数器数值（count）作为 CounterComp 组件的 Props 参数来实现就能满足基本需求了。不过这里要说明一下，常规的标准做法是使用 State（状态）来维护核心数据，这里采用 Props 参数也是可行的变通方法。另外，还需要在 CounterComp 组件中定义一组操作按钮，实现累加、累减、双倍和延时累加这几个功能，这些通过 React 框架就可以实现了。

下面是计数器应用视图模块（js 文件）的内容，具体代码如下：

【代码 12-3】（详见源代码 counter 目录下的 src\components\CounterComp.js 文件）

```js
01  import React, { Component } from 'react'
02  import PropTypes from 'prop-types'
03  import { CounterType } from '../actions'
04  // TODO: define ES6 Class React Component
05  class CounterComp extends Component {
06    // TODO: constructor
07    constructor(props) {
08      super(props);
09      this.onIncrement = this.onIncrement.bind(this);
10      this.onDecrement = this.onDecrement.bind(this);
11      this.onIncrementAsync = this.onIncrementAsync.bind(this);
12      this.onDouble = this.onDouble.bind(this);
13    }
14    // TODO: handle event
15    onIncrement() {
16      this.props.vstore.dispatch({ type: CounterType.INCREMENT });
17    }
18    onDecrement() {
19      this.props.vstore.dispatch({ type: CounterType.DECREMENT });
20    }
21    onDouble() {
22      this.props.vstore.dispatch({ type: CounterType.DOUBLE });
23    }
24    onIncrementAsync() {
25      setTimeout(this.onIncrement, 1000);
26    }
27    // TODO: render
28    render() {
29      // TODO: props
30      const value = this.props.value;
31      return (
32        <span>
33          <h3>基于 React + Redux 的计数器应用</h3>
34          <p>计数器：{value} 次.</p>
35          <p>
36            {' '}
37            <button onClick={this.onIncrement}> +1 </button>
38            {' '}
39            <button onClick={this.onDecrement}> -1 </button>
40            {' '}
41            <button onClick={this.onDouble}> Double </button>
42            {' '}
```

```
43              <button onClick={this.onIncrementAsync}> Increment Async </button>
44          </p>
45        </span>
46      );
47    }
48 }
49 // TODO: Props Type
50 CounterComp.propTypes = {
51   value: PropTypes.number.isRequired,
52   store: PropTypes.object
53 }
54 // TODO: export component
55 export default CounterComp;
```

关于【代码 12-3】的说明:

- 这段代码的大部分内容都是基于 React 框架实现的,如果读者掌握了前面章节的知识,对此就不难理解。下面着重介绍一下 React 之外的内容,其实 React 之外就是关于 Redux 的内容了。
- 第 30 行代码通过 Props 参数获取了一个 value 值,那该值是从何而来的呢? 先不着急,我们留在后面介绍。
- 第 34 行代码中直接使用该 value 值来展示计数器数值,这里读者即使不明白 value 值从何而来,也应该明白该 value 值维护了计数器应用的核心功能(即渲染计数器当前的数值)。
- 第 35～44 行代码定义了一组按钮,并添加了相应的单击事件处理方法。现在再看一下这几个单击事件处理方法的实现过程(见第 15～26 行代码),再次使用了 Props 参数获取了一个 vstore 属性,并调用了 dispatch 方法,在 dispatch 方法中发送了相应的 action 类型。
- 现在再把通过 Props 参数获取的两个属性(value 和 vstore)拿出来分析,我们会发现这两个属性来自于【代码 12-2】中第 13～14 行代码的定义。其中,value 属性用于保存计数器数值,vstore 属性用于保存 store 对象,然后通过 Props 参数的方式传递给 CounterComp 组件。
- 这样,CounterComp 组件就可以拿到 store 对象并通过维护计数器数值来实现渲染操作了,这就是该计数器应用的关键之处。

说　明
由于该计数器应用的功能相对简单,逻辑设计也相应地做了简化,标准做法是使用 State(状态)来维护核心数据的。

12.5 Action 定义

所谓 Action,就是把数据从应用(视图、服务器等)传递到 Store 的有效载荷,是 Store 数据的唯一来源。在 Redux 中,Action 是通过 Store 对象的 dispatch()方法传递到 Store 中的。

下面,我们看一下该计数器应用的 Action 定义,具体代码如下:

【代码 12-4】(详见源代码 counter 目录下的 src\actions\index.js 文件)

```
01  export const CounterType = {
02    INCREMENT: 'INCREMENT',
03    DECREMENT: 'DECREMENT',
04    DOUBLE: 'DOUBLE'
05  };
```

关于【代码 12-4】的说明:

- 这段代码很简单,定义了三种 action(INCREMENT、DECREMENT 和 DOUBLE),分别对应累加(+1)、累减(-1)和双倍(×2)。然后,直接通过 export 导出 action 类型(CounterType)。
- 在【代码 12-3】中,第 16 行、第 19 行和第 22 行代码通过 Store 对象的 dispatch()方法发送的 action,就是从 CounterType 获取的。

12.6 Reducer 设计

所谓 Reducer,就是指如何将应用 State(状态)的变化响应为 Action,并发送到 Store 的逻辑模块。请记住,Action 仅仅描述了事件发生的类型,并没有描述应用如何更新 State(状态)。对于具体更新 State(状态)的业务操作,是由 Reducer 负责处理完成的。

下面,我们看一下该计数器应用的 Reducer 设计,具体代码如下:

【代码 12-5】(详见源代码 counter 目录下的 src\reducers\reducerCounter.js 文件)

```
01  import { CounterType } from '../actions'
02  // TODO: reducer
03  var reducerCounter = (state = 0, action) => {
04    switch (action.type) {
05      case CounterType.INCREMENT:
06        return state + 1;
07      case CounterType.DECREMENT:
08        return state - 1;
```

```
09        case CounterType.DOUBLE:
10          return state * 2;
11        default:
12          return state;
13    }
14  };
15  // TODO: export Reducer
16  export default reducerCounter;
```

关于【代码12-5】的说明:

- Reducer 本质上就是一个纯函数,接收旧的 State(状态)和 Action,通过一定操作后返回新的 State(状态),具体说明如下:
  ➢ 第 01 行代码通过 import 引入了【代码12-4】定义的 Action。
  ➢ 第 03~14 行代码定义了一个 Reducer 处理函数(reducerCounter),传入了参数 state(旧的状态)和 action,然后第 04~13 行代码通过 action.type 处理分支逻辑,并操作返回新的状态。
- 第 16 行代码通过 export 导出 Reducer(reducerCounter)。

这里需要特别强调的是,保持 Reducer 函数的纯净很重要,切忌在 Reducer 里做以下这几类操作:

- 修改传入参数。
- 执行有副作用的操作,如 API 请求和路由跳转等。
- 调用非纯函数,如 Date.now()、Math.random()这类不确定返回值的函数。

12.7 计数器应用测试

在本节中,我们通过测试运行计数器应用,帮助读者进一步掌握基于 React + Redux 框架开发前端 Web 应用的过程。

通过 npm 工具启动计数器应用,具体命令如下:

```
npm start
```

如果应用启动成功,控制台中会打印出类似如图 12.3 所示的提示信息。

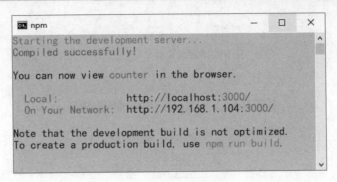

图 12.3　npm 控制台信息

同时，会启动默认浏览器打开应用页面（默认地址：http://localhost:3000），默认端口为 3000（可自定义修改），具体效果如图 12.4 所示。

图 12.4　计数器应用启动页面

下面，我们就可以在网页中通过操作按钮测试计数器应用的功能了，具体效果如图 12.5、图 12.6 和图 12.7 所示。

图 12.5　计数器应用（一）

图 12.6 计数器应用(二)

图 12.7 计数器应用(三)

如图 12.5、图 12.6 和图 12.7 中的箭头所示,通过点击不同的按钮(+1、Double、-1)触发了不同的 dispatch,然后根据 action 类型来对计数器数值状态进行操作,最后通过渲染操作将计数器数值状态变化体现在页面中。至于延时累加按钮(Increment Async)的操作,仅仅通过图片是无法展示效果的,读者可自行测试验证。

第 13 章

实战2：基于React+Redux实现计算器应用

本章介绍一个基于 React + Redux 实现的计算器应用，这是一个通过 Redux 结合 React 使用的最基本范例。

13.1 设计思想

本章介绍的这个计算器应用相对简易（模仿 Windows 系统自带的计算器），但对比于前一章实现的计数器应用还是更复杂一些。主要原因是需要设计一个计算器操作面板，导致需要 Redux 维护的 State（状态）数据有所增加。同时，还有一些数据是不需要 Redux 维护的，在设计架构时要先考虑清楚。

本计算器应用需要 Redux 维护的 State（状态）数据主要有计算结果、运算数按钮、运算符按钮和计算按钮等。在这些需要维护状态中，除了计算结果是数据类型外，其他的都属于操作类型。

为了便于读者更好地理解这个示例，我们根据计算器应用的主体架构绘制了一张逻辑图，如图 13.1 所示。

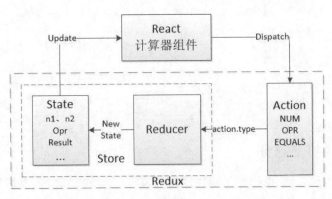

图 13.1　React + Redux 计算器应用逻辑图

图 13.1 中 Redux 模块中需要维护的 State（状态）主要有两个计算数、运算符和计算结果。相应地，action 主要定义了"NUM"（数值）、"OPR"（运算符）和"EQUALS"（计算结果）这几个类型。

关于计算器应用的代码结构如图 13.2 所示。

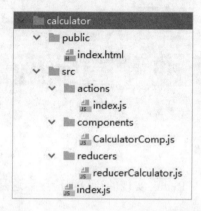

图 13.2 React + Redux 计算器应用结构图

图 13.2 中，文件目录名称（如 actions、components、reducers）基本代表了 js 模块文件的功能。

13.2 计算器应用页面

首先，看一下本章节实现的计算器应用的 HTML 主页，具体代码如下：

【代码 13-1】（详见源代码 calculator 目录下的 public\index.html 文件）

```
01  <!doctype html>
02  <html lang="en">
03    <head>
04      <meta charset="utf-8">
05      <meta name="viewport" content="width=device-width, initial-scale=1">
06      <title>Redux Calculator App</title>
07    </head>
08    <body>
09      <div id="root"></div>
10    </body>
11  </html>
```

关于【代码 13-1】的说明：

- 第 09 行代码通过<div id="root">标签定义的层元素渲染计算器组件。

13.3 主入口模块

在计算器应用的主入口模块中,主要是创建一个 Store 对象,作为 Redux 维护的数据,具体代码如下:

【代码 13-2】(详见源代码 calculator 目录下的 src\index.js 文件)

```
01  import React from 'react';
02  import ReactDOM from 'react-dom';
03  import { createStore } from 'redux';
04  import CalculatorComp from './components/CalculatorComp';
05  import reducerCalculator from './reducers/reducerCalculator';
06  // TODO: Create Store
07  const store = createStore(reducerCalculator);
08  const rootEle = document.getElementById('root');
09  // TODO: render
10  const render = () => ReactDOM.render(
11    <CalculatorComp
12      s={store.getState()}
13      vstore={store}
14    />,
15    rootEle
16  );
17  // TODO: call render
18  render();
19  // TODO: subscribe
20  store.subscribe(render);
```

关于【代码 13-2】的说明:

- 第 09 行代码通过调用 createStore()方法获取了 store 对象,参数是一个 Reducer 对象 (reducerCalculator)。
- 第 10~16 行代码通过调用 render()方法完成了页面渲染操作,具体说明如下:
  - ➢ 第 11~15 行代码通过自定义组件(CalculatorComp)实现了计算器。
  - ➢ 第 12 行代码定义了一个 value 属性,通过 store 对象的 getState()方法获取了计算器状态值。
  - ➢ 第 13 行代码定义了一个 vstore 属性,属性值为 store 对象。
- 第 20 行代码通过调用 store 对象的 subscribe()方法实现了订阅操作。

13.4 视图模块

在本应用中,计算器展示功能是通过 CalculatorComp 组件完成的。计算器 CalculatorComp 组件需要实现用户输入数字和运算符、完成运算和清除内容的功能,具体代码如下:

【代码 13-3】(详见源代码 calculator 目录下的 src\components\CalculatorComp.js 文件)

```
01  import React, {Component} from 'react'
02  import PropTypes from 'prop-types'
03  import {CalculateType} from '../actions'
04  // TODO: define ES6 Class React Component
05  class CalculatorComp extends Component {
06    // TODO: constructor
07    constructor(props) {
08      super(props);
09      this.state = {
10        b: 1
11      };
12      this.onNumClick = this.onNumClick.bind(this);
13      this.onOprClick = this.onOprClick.bind(this);
14      this.onEqualsClick = this.onEqualsClick.bind(this);
15      this.onClsClick = this.onClsClick.bind(this);
16    }
17    // TODO: handle event
18    onNumClick(e) {
19      e.preventDefault();
20      let n1, opr, n2, eq, result;
21      if (this.state.b) {
22        n1 = e.target.id.substr(3);
23        n2 = this.props.vstore.getState().n2;
24        opr = this.props.vstore.getState().opr;
25        eq = this.props.vstore.getState().eq;
26        result = this.props.vstore.getState().result;
27        console.log('n1: ' + n1);
28        console.log('n2: ' + n2);
29        this.setState({
30          b: 0
31        });
32      } else {
33        n1 = this.props.vstore.getState().n1;
34        n2 = e.target.id.substr(3);
35        opr = this.props.vstore.getState().opr;
```

```
36      eq = this.props.vstore.getState().eq;
37      result = this.props.vstore.getState().result;
38      console.log('n1: ' + n1);
39      console.log('n2: ' + n2);
40      this.setState({
41        b: 1
42      });
43    }
44    this.props.vstore.dispatch({
45      type: CalculateType.NUM,
46      n1: n1,
47      n2: n2,
48      opr: opr,
49      eq: eq,
50      result: result
51    });
52  }
53  onOprClick(e) {
54    e.preventDefault();
55    let opr = e.target.id.toString();
56    console.log('opr: ' + opr);
57    this.props.vstore.dispatch({
58      type: CalculateType.OPR,
59      n1: this.props.vstore.getState().n1,
60      n2: this.props.vstore.getState().n2,
61      opr: opr,
62      eq: this.props.vstore.getState().eq,
63      result: this.props.vstore.getState().result
64    });
65  }
66  onEqualsClick(e) {
67    e.preventDefault();
68    let eq = '=';
69    console.log('eq: ' + eq);
70    this.props.vstore.dispatch({
71      type: CalculateType.EQUALS,
72      n1: this.props.vstore.getState().n1,
73      n2: this.props.vstore.getState().n2,
74      opr: this.props.vstore.getState().opr,
75      eq: eq,
76      result: this.props.vstore.getState().result
77    });
78  }
```

```
79    onClsClick(e) {
80      e.preventDefault();
81      this.props.vstore.dispatch({
82        type: CalculateType.CLS,
83        n1: '',
84        n2: '',
85        opr: '',
86        eq: '',
87        result: ''
88      });
89    }
90    // TODO: render
91    render() {
92      // TODO: props
93      const n1 = this.props.s.n1;
94      const opr = this.props.s.opr;
95      const n2 = this.props.s.n2;
96      const eq = this.props.s.eq;
97      const result = this.props.s.result;
98      let lOpr;
99      switch (opr) {
100       case 'add':
101         lOpr = '+';
102         break;
103       case 'minus':
104         lOpr = '-';
105         break;
106       case 'multiple':
107         lOpr = HTMLDecode('&#215;');
108         console.log('multiple:' + lOpr);
109         break;
110       case 'divide':
111         lOpr = HTMLDecode('&#247;');
112         console.log('divide:' + lOpr);
113         break;
114       default:
115         lOpr = '';
116         break;
117     }
118     let expression = n1 + lOpr + n2 + eq + result;
119     return (
120       <span>
121         <h3>基于 React + Redux 的简单计算器应用</h3>
```

```
122            <p>计算结果: {expression}</p>
123            <p>
124            {' '}
125            <button id='num1' onClick={this.onNumClick}> 1 </button>
126            {' '}
127            <button id='num2' onClick={this.onNumClick}> 2 </button>
128            {' '}
129            <button id='num3' onClick={this.onNumClick}> 3 </button>
130            {' '}
131            <button id='add' onClick={this.onOprClick}> + </button><br/>
132            {' '}
133            <button id='num4' onClick={this.onNumClick}> 4 </button>
134            {' '}
135            <button id='num5' onClick={this.onNumClick}> 5 </button>
136            {' '}
137            <button id='num6' onClick={this.onNumClick}> 6 </button>
138            {' '}
139            <button id='minus' onClick={this.onOprClick}> - </button><br/>
140            {' '}
141            <button id='num7' onClick={this.onNumClick}> 7 </button>
142            {' '}
143            <button id='num8' onClick={this.onNumClick}> 8 </button>
144            {' '}
145            <button id='num9' onClick={this.onNumClick}> 9 </button>
146            {' '}
147            <button id='multiple' onClick={this.onOprClick}> &#215; </button><br/>
148            {' '}
149            <button id='num0' onClick={this.onNumClick}> 0 </button>
150            {' '}
151            <button id='cls' onClick={this.onClsClick}> C </button>
152            {' '}
153            <button id='equal' onClick={this.onEqualsClick}> = </button>
154            {' '}
155            <button id='divide' onClick={this.onOprClick}> &#247; </button><br/>
156            </p>
157         </span>
158      );
159   }
160 }
161 // TODO: Props Type
162 CalculatorComp.propTypes = {
```

```
163    result: PropTypes.string.isRequired,
164    store: PropTypes.object
165  }
166  // TODO: export component
167  export default CalculatorComp;
```

关于【代码13-3】的说明：

- 先看第 120~157 行代码定义的计算器界面，包括有数字按钮、运算符按钮和清除按钮，同时为按钮定义了相应的事件处理方法（onNumClick、onOprClick、onEqualsClick 和 onClsClick）。
- 第 18~89 行代码是以上事件处理方法（onNumClick、onOprClick、onEqualsClick 和 onClsClick）的实现过程，通过 Props 参数获取 Store 对象来调用 dispatch() 方法以发送 action 类型和状态参数（详见下一节内容）。
- 还有一点需要说明，第 10 行代码定义的 State（状态）数据（b），是用于区分第 1 个运算数和第 2 个运算数的。

13.5 Action 定义

本节介绍计算器应用的 Action 定义（CalculateType），具体代码如下：

【代码13-4】（详见源代码 calculator 目录下的 src\actions\index.js 文件）

```
01  export const CalculateType = {
02    NUM: 'NUM',
03    OPR: 'OPR',
04    EQUALS: 'EQUALS',
05    CLS: 'CLS'
06  };
```

关于【代码13-4】的说明：

- 第 02~05 行代码一共定义了 4 种 action（NUM、OPR、EQUALS 和 CLS），分别对应运算数、运算符、计算结果和清除。
- 在【代码13-3】中，第 45 行、第 58 行、第 71 行和第 82 行代码通过 Store 对象的 dispatch() 方法发送的 action，就是从 CalculateType 获取的。

13.6 Reducer 设计

本节介绍计算器应用中比较关键的 Reducer 设计，也就是具体业务处理的过程，具体代码如下：

【代码 13-5】（详见源代码 counter 目录下的 src\reducers\reducerCalculator.js 文件）

```
01  import { CalculateType } from '../actions'
02  // TODO: init state
03  const initState = {
04    n1: '',
05    n2: '',
06    opr: '',
07    eq: '',
08    result: ''
09  };
10  // TODO: reducer
11  var reducerCalculator = (state = initState, action) => {
12    switch (action.type) {
13      case CalculateType.NUM:
14        return {
15          n1: action.n1,
16          n2: action.n2,
17          opr: action.opr,
18          eq: action.eq,
19          result: action.result
20        };
21      case CalculateType.OPR:
22        return {
23          n1: action.n1,
24          n2: action.n2,
25          opr: action.opr,
26          eq: action.eq,
27          result: action.result
28        };
29      case CalculateType.EQUALS:
30        let result;
31        switch(action.opr) {
32          case 'add':
33            result = Number.parseInt(action.n1) + Number.parseInt(action.n2);
34            break;
35          case 'minus':
```

```
36            result = Number.parseInt(action.n1) - Number.parseInt(action.n2);
37            break;
38          case 'multiple':
39            result = Number.parseInt(action.n1) * Number.parseInt(action.n2);
40            break;
41          case 'divide':
42            if(Number.parseInt(action.n2) !== 0) {
43              result = Number.parseInt(action.n1) / Number.parseInt(action.n2);
44            } else {
45              result = Number.NaN;
46            }
47            break;
48          default:
49            result = 0;
50            break;
51        }
52        return {
53          n1: action.n1,
54          n2: action.n2,
55          opr: action.opr,
56          eq: action.eq,
57          result: result
58        };
59      case CalculateType.CLS:
60        return {
61          n1: action.n1,
62          n2: action.n2,
63          opr: action.opr,
64          eq: action.eq,
65          result: action.result
66        };
67      default:
68        return state;
69    }
70  };
71  // TODO: export Reducer
72  export default reducerCalculator;
```

关于【代码 13-5】的说明：

- Reducer 的主要功能就是接收旧的 State（状态）和 Action，通过具体操作后返回新的 State（状态），具体说明如下：
  - 第 03~09 行代码定义了一个初始化状态（initState），将作为参数传递给下面定义的 Reducer 处理函数（reducerCalculator）。
- 第 11~70 行代码定义了一个 Reducer 处理函数（reducerCalculator），传入了参数 state（初始化状态 initState）和 action，然后通过 action.type 处理分支逻辑并操作返回新的状态。其中，第 31~51 行代码就是具体的"加减乘除"运算过程。
- 第 72 行代码通过 export 导出 Reducer（reducerCalculator）。

13.7 计算器应用测试

本节我们测试运行一下计算器应用，页面初始效果如图 13.3 所示。

图 13.3　计算器应用初始页面

下面就可以在网页中通过操作按钮测试计算器应用的功能了，具体效果如图 13.4、图 13.5、图 13.6 和图 13.7 所示。

图 13.4　计算器应用（一）

图 13.5　计算器应用（二）

图 13.6　计算器应用（三）

图 13.7　计算器应用（四）

图 13.4 至图 13.7 所示为"加减乘除"四种运算，结果均正确无误。

第 14 章

实战3：基于Provider容器组件重构计算器应用

本章介绍基于 React-Redux 库中的 Provider 容器组件重构计算器应用的方法，这是一个目前比较流行的做法。

14.1 设计思想

我们回顾前一章所实现的计算器应用的主入口模块，具体代码如下：

【代码 14-1】（节选自【代码 13-2】）

```
11    <CalculatorComp
12      s={store.getState()}
13      vstore={store}
14    />
```

关于【代码 14-1】的说明：

- 这段代码引用了计算器组件（CalculatorComp），同时定义了属性 s（State 状态）和属性 vstore（Store 对象），这两个属性将作为 Props 参数传递给计算器组件（CalculatorComp）。

不过【代码 14-1】的方式仅仅适用于属性较少的情况，如果需要一大堆属性作为 Props 参数进行传递，那性能成本就无法估量了。

因此，React-Redux 库（Redux 提供的 React 官方绑定库）设计了一个 Provider 容器组件来解决这个问题，同时还提供了一个 connect()方法来连接组件于 Store 对象。关于 connect()方法的语法描述如下：

```
connect(mapStateToProps, mapDispatchToProps, mergeProps, options = {})
```

这里就不针对语法进行具体解释了，我们将通过后面的实例代码帮助读者学习使用该方法。

为了便于读者更好地理解这个示例,我们根据重构的计算器应用主体架构绘制了一张逻辑图,如图 14.1 所示。

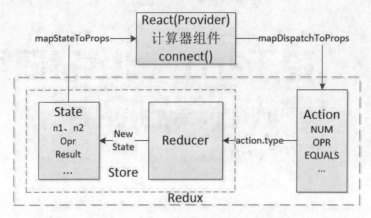

图 14.1　React + Redux 计算器应用逻辑图

上图中,Provider 计算器组件将 Store 对象包装在顶层容器中,这样就可以被其子组件继承使用了。

这个基于 Provider 容器组件重构的计算器应用的代码结构如图 14.2 所示。

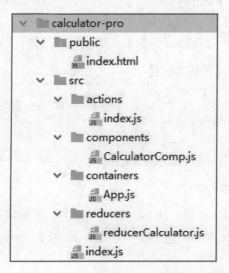

图 14.2　Provider 容器组件计算器应用结构图

图 14.2 中,文件目录名称(如 actions、components、containers、reducers)基本代表了 js 模块文件的功能。其中,新增的 containers 表示容器组件目录。

14.2 主入口模块

对于重构的计算器应用的主入口模块,创建 Store 对象的方式是一致的,主要变化是引入了 Provider 容器组件,具体代码如下:

【代码 14-2】(详见源代码 calculator-pro 目录下的 src\index.js 文件)

```
01  import React from 'react';
02  import ReactDOM from 'react-dom';
03  import { createStore } from 'redux';
04  import { Provider } from 'react-redux';
05  import reducerCalculator from './reducers/reducerCalculator';
06  import { App } from './containers/App';
07  // TODO: Create Store
08  const store = createStore(reducerCalculator);
09  const rootEle = document.getElementById('root');
10  // TODO: render
11  const render = () => ReactDOM.render(
12    <Provider store={store}>
13      <App />
14    </Provider>,
15  rootEle
16  );
17  // TODO: call render
18  render();
19  // TODO: subscribe
20  store.subscribe(render);
```

关于【代码 14-2】的说明:

- 第 12~14 行代码通过<Provider>定义了一个容器组件,同时添加了 store 属性。
- 第 13 行代码在<Provider>容器组件中包裹了一个<App>模块组件。

这段代码的核心就是对于 Provider 容器组件的使用,store 对象作为<Provider>的属性使用,并向下传递给子组件。

14.3 App 组件

在【代码 14-2】中,第 13 行代码引用的<App>模块组件是负责管理 State(状态)和 dispatch 方法的,具体方式就是通过第一节介绍的 connect()方法来实现。下面看一下具体的代码:

【代码 14-3】（详见源代码 calculator-pro 目录下的 src\containers\App.js 文件）

```
01  import { connect } from 'react-redux';
02  import { CalculateType } from '../actions';
03  import CalculatorComp from '../components/CalculatorComp';
04  // TODO: mapStateToProps
05  function mapStateToProps(state) {
06    return state;
07  }
08  // TODO: mapDispatchToProps
09  function mapDispatchToProps(dispatch) {
10    return {
11      onNumClick: (n1, n2, opr, eq, result) => dispatch({
12        type: CalculateType.NUM,
13        n1: n1,
14        n2: n2,
15        opr: opr,
16        eq: eq,
17        result: result
18      }),
19      onOprClick: (n1, n2, opr, eq, result) => dispatch({
20        type: CalculateType.OPR,
21        n1: n1,
22        n2: n2,
23        opr: opr,
24        eq: eq,
25        result: result
26      }),
27      onEqualsClick: (n1, n2, opr, eq, result) => dispatch({
28        type: CalculateType.EQUALS,
29        n1: n1,
30        n2: n2,
31        opr: opr,
32        eq: eq,
33        result: result
34      }),
35      onClsClick: (n1, n2, opr, eq, result) => dispatch({
36        type: CalculateType.CLS,
37        n1: n1,
38        n2: n2,
39        opr: opr,
40        eq: eq,
41        result: result
42      })
```

```
43    }
44  }
45  // TODO: export
46  export const App = connect(
47    mapStateToProps,
48    mapDispatchToProps
49  )(CalculatorComp);
```

关于【代码 14-3】的说明：

- 第 05~07 行代码定义的 mapStateToProps()方法，负责将需要管理的 State（状态）映射到 Props 参数上。
- 第 09~44 行代码定义的 mapDispatchToProps()方法，负责将一组 dispatch()方法映射到 Props 参数上。这组 dispatch()方法负责管理 State（状态）和 action 类型，与【代码 13-3】对比仅仅是形式不同，功能完全一致。
- 第 46~49 行代码通过调用 connect()方法，先定义参数为 mapStateToProps 和 mapDispatchToProps，然后关联到视图组件（CalculatorComp）上。

14.4 视图模块

重构后的视图模块基本功能是一致的，但是在具体方式上要进行相应的变动，具体代码如下：

【代码 14-4】（详见源代码 calculator-pro 目录下的 src\components\CalculatorComp.js）

```
01  import React, {Component} from 'react'
02  import PropTypes from 'prop-types'
03  // TODO: define ES6 Class React Component
04  class CalculatorComp extends Component {
05    // TODO: constructor
06    constructor(props) {
07      super(props);
08      console.log("props:", props);
09      this.state = {
10        b: 1
11      };
12      this.onNumClick = this.onNumClick.bind(this);
13      this.onOprClick = this.onOprClick.bind(this);
14      this.onEqualsClick = this.onEqualsClick.bind(this);
15      this.onClsClick = this.onClsClick.bind(this);
16    }
```

```
17      // TODO: handle event
18      onNumClick(e) {
19        e.preventDefault();
20        let num = e.target.id.substr(3);
21        console.log('num: ' + num);
22        if (this.state.b) {
23          this.props.onNumClick(num,this.props.n2,this.props.opr,this.props.eq,this.props.result);
24          this.setState({
25            b: 0
26          });
27        } else {
28          this.props.onNumClick(this.props.n1,num,this.props.opr,this.props.eq,this.props.result);
29          this.setState({
30            b: 1
31          });
32        }
33      }
34      onOprClick(e) {
35        e.preventDefault();
36        let opr = e.target.id.toString();
37        console.log('opr: ' + opr);
38        this.props.onOprClick(this.props.n1, this.props.n2, opr, this.props.eq,this.props.result);
39      }
40      onEqualsClick(e) {
41        e.preventDefault();
42        let eq = '=';
43        console.log('eq: ' + eq);
44        console.log("props:", this.props);
45      this.props.onEqualsClick(this.props.n1, this.props.n2, this.props.opr, eq, this.props.result);
46      }
47      onClsClick(e) {
48        e.preventDefault();
49  this.props.onClsClick(this.props.n1,this.props.n2,this.props.opr,this.props.eq,this.props.result);
50      }
51      // TODO: render
52      render() {
53        // TODO: props
```

```
54      console.log("props:", this.props);
55      const {n1,n2,opr,eq,result,onNumClick,onOprClick,onEqualsClick,onClsClick} = this.props;
56      let lOpr;
57      switch (opr) {
58        case 'add':
59          lOpr = '+';
60          break;
61        case 'minus':
62          lOpr = '-';
63          break;
64        case 'multiple':
65          lOpr = HTMLDecode('&#215;');
66          console.log('multiple:' + lOpr);
67          break;
68        case 'divide':
69          lOpr = HTMLDecode('&#247;');
70          console.log('divide:' + lOpr);
71          break;
72        default:
73          lOpr = '';
74          break;
75      }
76      let expression = n1 + lOpr + n2 + eq + result;
77      return (
78        <span>
79        <h3>基于 Provider 的重构计算器应用</h3>
80        <p>计算结果：{expression}</p>
81        <p>
82        {' '}
83        <button id='num1' onClick={this.onNumClick}> 1 </button>
84        {' '}
85        <button id='num2' onClick={this.onNumClick}> 2 </button>
86        {' '}
87        <button id='num3' onClick={this.onNumClick}> 3 </button>
88        {' '}
89        <button id='add' onClick={this.onOprClick}> + </button><br/>
90        {' '}
91        <button id='num4' onClick={this.onNumClick}> 4 </button>
92        {' '}
93        <button id='num5' onClick={this.onNumClick}> 5 </button>
94        {' '}
```

```
 95        <button id='num6' onClick={this.onNumClick}> 6 </button>
 96        {' '}
 97        <button id='minus' onClick={this.onOprClick}> - </button><br/>
 98        {' '}
 99        <button id='num7' onClick={this.onNumClick}> 7 </button>
100        {' '}
101        <button id='num8' onClick={this.onNumClick}> 8 </button>
102        {' '}
103        <button id='num9' onClick={this.onNumClick}> 9 </button>
104        {' '}
105        <button id='multiple' onClick={this.onOprClick}> &#215; </button><br/>
106        {' '}
107        <button id='num0' onClick={this.onNumClick}> 0 </button>
108        {' '}
109        <button id='cls' onClick={this.onClsClick}> C </button>
110        {' '}
111        <button id='equal' onClick={this.onEqualsClick}> = </button>
112        {' '}
113        <button id='divide' onClick={this.onOprClick}> &#247; </button><br/>
114       </p>
115      </span>
116    );
117   }
118 }
119 // TODO: Props Type
120 CalculatorComp.propTypes = {
121   result: PropTypes.number.isRequired,
122   store: PropTypes.object
123 }
124 // TODO: export component
125 export default CalculatorComp;
```

关于【代码 14-4】的说明：

- 第 18～33 行代码、第 34～39 行代码、第 40～46 行代码和第 47～50 行代码定义了一组事件处理方法，在这些事件处理方法对应于 mapDispatchToProps() 方法（见【代码 14-3】）定义的一组 dispatch 方法。

- 第 55 行代码通过 Props 参数获取了 State（状态）值和事件方法，然后放在 render() 方法中使用。

14.5 Action 定义

本节介绍重构的计算器应用的 Action 定义（CalculateType），具体代码如下：

【代码 14-5】（详见源代码 calculator-pro 目录下的 src\actions\index.js 文件）

```
01  export const CalculateType = {
02    NUM: 'NUM',
03    OPR: 'OPR',
04    EQUALS: 'EQUALS',
05    CLS: 'CLS'
06  };
```

关于【代码 14-5】的说明：

- 这段代码与【代码 13-5】相同，说明 Action 类型的定义是完全一致的。

14.6 Reducer 设计

本节介绍重构的计算器应用中比较关键的 Reducer 设计，也就是具体业务处理的过程，具体代码如下：

【代码 14-6】（详见源代码 counter 目录下的 src\reducers\reducerCalculator.js 文件）

```
01  import { CalculateType } from '../actions'
02  // TODO: init state
03  const initState = {
04    n1: '',
05    n2: '',
06    opr: '',
07    eq: '',
08    result: ''
09  };
10  // TODO: reducer
11  var reducerCalculator = (state = initState, action) => {
12    switch (action.type) {
13      case CalculateType.NUM:
14        return {
15          n1: action.n1,
16          n2: action.n2,
17          opr: action.opr,
18          eq: action.eq,
```

```
19          result: action.result
20        };
21    case CalculateType.OPR:
22      return {
23        n1: action.n1,
24        n2: action.n2,
25        opr: action.opr,
26        eq: action.eq,
27        result: action.result
28      };
29    case CalculateType.EQUALS:
30      let result;
31      console.log('action:' + action);
32      switch(action.opr) {
33        case 'add':
34          return {
35            n1: action.n1,
36            n2: action.n2,
37            opr: action.opr,
38            eq: action.eq,
39            result: Number.parseInt(action.n1) + Number.parseInt(action.n2)
40          };
41        case 'minus':
42          return {
43            n1: action.n1,
44            n2: action.n2,
45            opr: action.opr,
46            eq: action.eq,
47            result: Number.parseInt(action.n1) - Number.parseInt(action.n2)
48          };
49        case 'multiple':
50          return {
51            n1: action.n1,
52            n2: action.n2,
53            opr: action.opr,
54            eq: action.eq,
55            result: Number.parseInt(action.n1) * Number.parseInt(action.n2)
56          };
57        case 'divide':
58          if(Number.parseInt(action.n2) !== 0) {
59            return {
60              n1: action.n1,
```

```
61              n2: action.n2,
62              opr: action.opr,
63              eq: action.eq,
64              result: Number.parseInt(action.n1) / Number.parseInt(action.n2)
65            }
66          } else {
67            return {
68              n1: action.n1,
69              n2: action.n2,
70              opr: action.opr,
71              eq: action.eq,
72              result: Number.NaN
73            }
74          }
75        default:
76          return {
77            n1: action.n1,
78            n2: action.n2,
79            opr: action.opr,
80            eq: action.eq,
81            result: 0
82          }
83      }
84    case CalculateType.CLS:
85      return {
86        n1: '',
87        n2: '',
88        opr: '',
89        eq: '',
90        result: ''
91      };
92    default:
93      return state;
94  }
95 };
96 // TODO: export Reducer
97 export default reducerCalculator;
```

关于【代码14-6】的说明：

- Reducer的主要功能就是接收旧的State（状态）和Action，通过操作后返回新的State（状态），具体说明如下：
  ➢ 第03~09行代码定义了一个初始化状态（initState），将作为参数传递给下面定义的 Reducer 处理函数（reducerCalculator）。
➢ 第 11～95 行代码定义了一个 Reducer 处理函数（reducerCalculator），传入了参数 state（初始化状态 initState）和 action，然后通过 action.type 处理分支逻辑，并操作返回新的状态。其中，第 32～83 行代码就是具体的"加减乘除"运算过程。
➢ 第 97 行代码通过 export 导出 Reducer（reducerCalculator）。

14.7 重构的计算器应用测试

下面，我们测试运行一下重构的计算器应用，页面效果如图 14.3 所示。

图 14.3　重构计算器应用页面

具体的计算测试这里就不给出了，与前一章的效果基本一致。

第 15 章

实战4：基于Redux实现任务管理器应用

本章介绍一个基于 React + Redux 实现的任务管理器应用，这是一个通过 Redux 结合 React 使用的高级应用范例。

15.1 设计思想

本章介绍的这个任务管理器应用比较简单，主要是设计一个任务管理的操作面板，可以满足新建任务、激活任务和禁止任务等几个需求。

为了便于读者更好地理解这个示例，我们根据任务管理器应用的主体架构绘制了一张逻辑图，如图 15.1 所示。

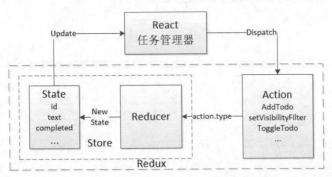

图 15.1　任务管理器应用逻辑图

在图 15.1 中，Redux 模块中需要维护的 State（状态）主要有 id、text 和 completed。相应地，Action 主要定义了"AddTodo""setVisibilityFilter"和"ToggleTodo"这几个类型。

任务管理器应用的代码结构如图 15.2 所示。

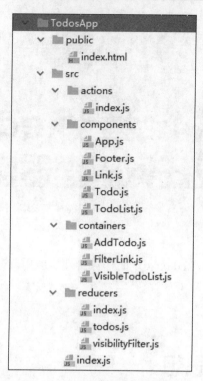

图 15.2　任务管理器应用结构图

15.2 任务管理器应用页面

首先，看一下任务管理器应用的 HTML 主页，具体代码如下：

【代码 15-1】（详见源代码 todosApp 目录下的 public\index.html 文件）

```
01  <!doctype html>
02  <html lang="en">
03    <head>
04      <meta charset="utf-8">
05      <meta name="viewport" content="width=device-width, initial-scale=1">
06      <title>Redux TodosApp Example</title>
07    </head>
08    <body>
09      <div id="root"></div>
10    </body>
11  </html>
```

关于【代码 15-1】的说明：

- 第 09 行代码通过<div id="root">标签定义的层元素用于渲染任务管理器组件。

15.3　主入口模块

对于任务管理器应用的主入口模块，首先还是创建 store 对象，然后通过 Provider 容器组件方式引入 Store 对象并向下传递，具体代码如下：

【代码 15-2】（详见源代码 todosApp 目录下的 src\index.js 文件）

```
01  import React from 'react';
02  import { render } from 'react-dom';
03  import { createStore } from 'redux';
04  import { Provider } from 'react-redux';
05  import App from './components/App';
06  import rootReducer from './reducers';
07  // TODO: Create Store Obj
08  const store = createStore(rootReducer);
09  // TODO: render
10  render(
11    <Provider store={store}>
12      <App />
13    </Provider>,
14    document.getElementById('root')
15  );
```

关于【代码 15-2】的说明：

- 第 11～13 行代码通过<Provider>定义了一个容器组件，同时添加了 store 属性。
- 第 12 行代码在<Provider>容器组件中包裹了一个<App>模块组件。

15.4　App 组件

在【代码 15-2】中，第 12 行代码引用的<App>模块组件是负责定义页面组件的，具体代码如下：

【代码 15-3】（详见源代码 todosApp 目录下的 src\components\App.js 文件）

```
01  import React from 'react';
02  import Footer from './Footer';
03  import AddTodo from '../containers/AddTodo';
04  import VisibleTodoList from '../containers/VisibleTodoList';
05  // TODO: App
06  const App = () => (
```

```
07      <div>
08          <AddTodo />
09          <VisibleTodoList />
10          <Footer />
11      </div>
12  );
13  // TODO: export App
14  export default App;
```

关于【代码 15-3】的说明：

- 第 06~12 行代码定义的 App 组件，包含了 3 个子组件，具体说明如下：
  ➢ 第 08 行代码引入了一个<AddTodo>子组件，用于新增任务项。
  ➢ 第 09 行代码引入了一个<VisibleTodoList>子组件，用于管理全部任务项列表。
  ➢ 第 10 行代码引入了一个<Footer>子组件，用于定义任务筛选功能。
- 第 14 行代码通过 export 导出 App 组件。

15.5 AddTodo 组件

在【代码 15-3】中，第 08 行代码引用的<AddTodo>组件负责新增任务项，具体代码如下：

【代码 15-4】（详见源代码 todosApp 目录下的 src\containers\AddTodo.js 文件）

```
01  import React from 'react';
02  import { connect } from 'react-redux';
03  import { addTodo } from '../actions';
04  // TODO: AddTodo
05  const AddTodo = ({ dispatch }) => {
06      let input;
07      return (
08          <div>
09              <h3>基于 Redux 实现任务管理器</h3>
10              <form onSubmit={e => {
11                  e.preventDefault();
12                  if (!input.value.trim()) {
13                      return
14                  }
15                  dispatch(addTodo(input.value));
16                  input.value = ''
17              }}>
18                  <input ref={node => input = node} />
19                  {' '}
```

```
20        <button type="submit">
21          新增任务项
22        </button>
23      </form>
24    </div>
25  );
26 }
27 // TODO: export
28 export default connect()(AddTodo);
```

关于【代码 15-4】的说明：

- 第 05～26 行代码定义的 AddTodo 组件，具体说明如下：
  - 第 06 行代码定义了一个变量（input），用于保存用户输入的任务项名称。
  - 第 10～23 行代码定义了一个表单，用于接收用户输入的任务项名称，然后执行提交操作。
- 第 28 行代码通过调用 connect()方法关联 AddTodo 组件并导出该组件。

15.6 VisibleTodoList 组件

在【代码 15-3】中，第 09 行代码引用的<VisibleTodoList>组件管理全部任务项列表，具体代码如下：

【代码 15-5】（详见源代码 todosApp 目录下的 src\containers\VisibleTodoList.js 文件）

```
01 import { connect } from 'react-redux';
02 import { toggleTodo } from '../actions';
03 import TodoList from '../components/TodoList';
04 import { VisibilityFilters } from '../actions';
05 // TODO: getVisibleTodos
06 const getVisibleTodos = (todos, filter) => {
07   switch (filter) {
08     case VisibilityFilters.SHOW_ALL:
09       return todos;
10     case VisibilityFilters.SHOW_COMPLETED:
11       return todos.filter(t => t.completed);
12     case VisibilityFilters.SHOW_ACTIVE:
13       return todos.filter(t => !t.completed);
14     default:
15       throw new Error('Unknown filter: ' + filter);
16   }
17 }
```

```
18  // TODO: mapStateToProps
19  const mapStateToProps = state => ({
20    todos: getVisibleTodos(state.todos, state.visibilityFilter)
21  });
22  // TODO: mapDispatchToProps
23  const mapDispatchToProps = dispatch => ({
24    toggleTodo: id => dispatch(toggleTodo(id))
25  });
26  // TODO: export
27  export default connect(
28    mapStateToProps,
29    mapDispatchToProps
30  )(TodoList);
```

关于【代码 15-5】的说明：

- 第 19~21 行代码定义的 mapStateToProps()方法，负责将需要管理的 State（状态）映射到 Props 参数上。其中，第 20 行代码通过调用自定义 getVisibleTodos()方法筛选 action。
- 第 06~17 行代码是自定义 getVisibleTodos()方法的具体实现过程。
- 第 23~25 行代码定义的 mapDispatchToProps()方法，负责将 dispatch()方法映射到 Props 参数上。
- 第 27~30 行代码通过调用 connect()方法关联 TodoList 组件并导出该组件。

下面再看一下关于 TodoList 组件的代码：

【代码 15-6】（详见源代码 todosApp 目录下的 src\components\TodoList.js 文件）

```
01  import React from 'react';
02  import PropTypes from 'prop-types';
03  import Todo from './Todo';
04  // TODO: TodoList
05  const TodoList = ({ todos, toggleTodo }) => (
06    <ul>
07      {todos.map(todo =>
08        <Todo
09          key={todo.id}
10          {...todo}
11          onClick={() => toggleTodo(todo.id)}
12        />
13      )}
14    </ul>
15  )
16  // TODO: Props Types
```

```
17  TodoList.propTypes = {
18    todos: PropTypes.arrayOf(PropTypes.shape({
19      id: PropTypes.number.isRequired,
20      completed: PropTypes.bool.isRequired,
21      text: PropTypes.string.isRequired
22    }).isRequired).isRequired,
23    toggleTodo: PropTypes.func.isRequired
24  }
25  // TODO: export
26  export default TodoList;
```

关于【代码 15-6】的说明：

- 第 05～15 行代码定义了一个 TodoList 组件负责任务列表的显示，具体说明如下：
  > 第 07～13 行代码通过调用 map()映射方法创建列表项。
  > 第 08～12 行代码通过引入<Todo>子组件创建具体的列表项。
- 第 26 行代码通过 export 导出 TodoList 组件。

然后，再看一下关于 Todo 组件的代码：

【代码 15-7】（详见源代码 todosApp 目录下的 src\components\Todo.js 文件）

```
01  import React from 'react';
02  import PropTypes from 'prop-types';
03  // TODO: Todo
04  const Todo = ({ onClick, completed, text }) => (
05    <li
06      onClick={onClick}
07      style={{
08        textDecoration: completed ? 'line-through' : 'none'
09      }}
10    >
11      {text}
12    </li>
13  );
14  // TODO: Props Types
15  Todo.propTypes = {
16    onClick: PropTypes.func.isRequired,
17    completed: PropTypes.bool.isRequired,
18    text: PropTypes.string.isRequired
19  }
20  // TODO: export
21  export default Todo;
```

关于【代码 15-7】的说明：

- 第 04～13 行代码定义了一个 Todo 组件负责创建任务列表项，具体说明如下：
  - 第 05～12 行代码定义了一个<li>列表项，并添加了 onClick、style 等属性。
  - 第 06 行代码定义的 onClick 属性负责任务列表项单击事件的处理。
  - 第 11 行代码定义的 text 属性负责显示任务列表项名称。
- 第 21 行代码通过 export 导出 Todo 组件。

15.7 Footer 组件

在【代码 15-3】中，第 10 行代码引用的<Footer>组件任务筛选功能，具体代码如下：

【代码 15-8】（详见源代码 todosApp 目录下的 src\components\Footer.js 文件）

```
01  import React from 'react';
02  import FilterLink from '../containers/FilterLink';
03  import { VisibilityFilters } from '../actions';
04  // TODO: Footer
05  const Footer = () => (
06    <div>
07      <span>筛选显示：</span>
08      <FilterLink filter={VisibilityFilters.SHOW_ALL}>
09        全部
10      </FilterLink>
11      <FilterLink filter={VisibilityFilters.SHOW_ACTIVE}>
12        激活
13      </FilterLink>
14      <FilterLink filter={VisibilityFilters.SHOW_COMPLETED}>
15        禁用
16      </FilterLink>
17    </div>
18  )
19  // TODO: export
20  export default Footer;
```

关于【代码 15-8】的说明：

- 第 05～18 行代码定义了一个 Footer 组件，其中第 08 行、第 11 行和第 14 行代码通过引入<FilterLink>定义了一组筛选操作按钮。
- 第 20 行代码通过 export 导出了 Footer 组件。

下面再看一下关于 FilterLink 组件的代码：

第 15 章 实战 4：基于 Redux 实现任务管理器应用

【代码 15-9】（详见源代码 todosApp 目录下的 src\containers\FilterLink.js 文件）

```
01  import { connect } from 'react-redux';
02  import { setVisibilityFilter } from '../actions';
03  import Link from '../components/Link';
04  // TODO: mapStateToProps
05  const mapStateToProps = (state, ownProps) => ({
06    active: ownProps.filter === state.visibilityFilter
07  });
08  // TODO: mapDispatchToProps
09  const mapDispatchToProps = (dispatch, ownProps) => ({
10    onClick: () => dispatch(setVisibilityFilter(ownProps.filter))
11  });
12  // TODO: export
13  export default connect(
14    mapStateToProps,
15    mapDispatchToProps
16  )(Link);
```

关于【代码 15-9】的说明：

- 第 05~07 行代码定义的 mapStateToProps()方法，负责将需要管理的 State（状态）映射到 Props 参数上。
- 第 09~11 行代码定义的 mapDispatchToProps()方法，负责将 dispatch()方法映射到 Props 参数上。
- 第 13~16 行代码通过调用 connect()方法关联 Link 组件并导出该组件。

然后，再看一下关于 Link 组件的代码：

【代码 15-10】（详见源代码 todosApp 目录下的 src\components\Link.js 文件）

```
01  import React from 'react';
02  import PropTypes from 'prop-types';
03  // TODO: Link
04  const Link = ({ active, children, onClick }) => (
05    <button
06      onClick={onClick}
07      disabled={active}
08      style={{
09        marginLeft: '4px'
10      }}
11    >
12      {children}
13    </button>
14  );
15  // TODO: Props Types
```

```
16  Link.propTypes = {
17    active: PropTypes.bool.isRequired,
18    children: PropTypes.node.isRequired,
19    onClick: PropTypes.func.isRequired
20  }
21  // TODO: export
22  export default Link;
```

关于【代码 15-10】的说明：

- 第 04~14 行代码定义了一个 Link 组件负责任务列表的管理，具体说明如下：
  - 第 05~13 行代码定义了一个<button>按钮，并添加了 onClick、disabled、style 等属性。
  - 第 06 行代码定义的 onClick 属性负责任务列表项单击事件的处理。
  - 第 07 行代码定义的 disabled 属性负责任务列表项激活状态与禁止状态的切换。
- 第 22 行代码通过 export 导出 Link 组件。

15.8 Action 定义

本节介绍任务管理器应用的 Action 定义，具体代码如下：

【代码 15-11】（详见源代码 todosApp 目录下的 src\actions\index.js 文件）

```
01  let nextTodoId = 0;
02  export const addTodo = text => ({
03    type: 'ADD_TODO',
04    id: nextTodoId++,
05    text
06  });
07  export const setVisibilityFilter = filter => ({
08    type: 'SET_VISIBILITY_FILTER',
09    filter
10  });
11  export const toggleTodo = id => ({
12    type: 'TOGGLE_TODO',
13    id
14  });
15  export const VisibilityFilters = {
16    SHOW_ALL: 'SHOW_ALL',
17    SHOW_COMPLETED: 'SHOW_COMPLETED',
18    SHOW_ACTIVE: 'SHOW_ACTIVE'
19  }
```

关于【代码 15-11】的说明：

- 第 01 行代码定义了一个变量（nextTodoId），用于保存 id 属性值。
- 第 02~06 行代码定义的 addTodo 方法负责导出 action 类型（ADD_TODO）。
- 第 07~10 行代码定义的 setVisibilityFilter 方法负责导出 action 类型（SET_VISIBILITY_FILTER）。
- 第 11~14 行代码定义的 toggleTodo 方法负责导出 action 类型（TOGGLE_TODO）。
- 第 15~19 行代码定义的 VisibilityFilters 方法负责导出一组 action 类型（SHOW_ALL、SHOW_COMPLETED 和 SHOW_ACTIVE）。

15.9 Reducer 设计

本节介绍任务管理器应用中比较关键的 Reducer 设计，也就是具体业务处理的过程。对于任务管理器应用的复杂程度，我们可以考虑将 Reducer 函数拆分成多个单独的函数，拆分后的每个函数独立负责管理 State（状态）的一部分。

Redux 为该设计方式提供了一个辅助函数（combineReducers），该函数的作用就是把一个多个不同的 Reducer 函数（单独管理不同的 State）合并成一个最终的 Reducer 函数，然后就可以对这个 Reducer 函数调用 createStore 方法。

通过 combineReducers() 方法合并后的 Reducer 函数可以调用各个子 Reducer 函数，并将返回结果合并成一个 State 对象。该方法返回的 State 对象，会将传入的每个 reducer 返回的 State 对象，按其传递给 combineReducers() 方法时所对应的 key 进行命名。

下面看一下具体的代码：

【代码 15-12】（详见源代码 counter 目录下的 src\reducers\index.js 文件）

```
01  import { combineReducers } from 'redux';
02  import todos from './todos';
03  import visibilityFilter from './visibilityFilter';
04  // TODO: export
05  export default combineReducers({
06    todos,
07    visibilityFilter
08  });
```

关于【代码 15-12】的说明：

- 第 01 行代码从 redux 引入了 combineReducers 方法。
- 第 02、03 行代码分别引入了两个子模块（todos 和 visibilityFilter）。
- 第 05~08 行代码通过 combineReducers() 方法包裹了上面引入的两个子模块（todos 和 visibilityFilter），也就是将两个独立的 Reducer 合并成一个 Reducer，然后通过 export 导出该 Reducer。

下面再看一下关于 todos 模块的代码：

【代码 15-13】（详见源代码 todosApp 目录下的 src\reducers\todos.js 文件）

```
01  // TODO: Reducer - todos
02  const todos = (state = [], action) => {
03    switch (action.type) {
04      case 'ADD_TODO':
05        return [
06          ...state,
07          {
08            id: action.id,
09            text: action.text,
10            completed: false
11          }
12        ]
13      case 'TOGGLE_TODO':
14        return state.map(todo =>
15          (todo.id === action.id)
16            ? {...todo, completed: !todo.completed}
17            : todo
18        )
19      default:
20        return state
21    }
22  }
23  // TODO: export
24  export default todos;
```

关于【代码 15-13】的说明：

- 第 02 ~ 22 行代码定义了一个 Reducer 函数（todos），通过判断 action 类型（ADD_TODO 和 TOGGLE_TODO）来执行管理任务列表项的操作。
- 第 24 行代码通过 export 导出了 todos 模块。

然后，再看一下关于 visibilityFilter 模块的代码：

【代码 15-14】（详见源代码 todosApp 目录下的 src\reducers\visibilityFilter.js 文件）

```
01  import { VisibilityFilters } from '../actions';
02  // TODO: visibilityFilter
03  const visibilityFilter = (state=VisibilityFilters.SHOW_ALL, action) => {
04    switch (action.type) {
05      case 'SET_VISIBILITY_FILTER':
06        return action.filter;
07      default:
08        return state;
09    }
```

```
10  }
11  // TODO: export
12  export default visibilityFilter;
```

关于【代码 15-14】的说明：

- 第 03～10 行代码定义了一个 Reducer 函数（visibilityFilter），通过判断 action 类型（SET_VISIBILITY_FILTER）来执行筛选任务列表项的操作。
- 第 12 行代码通过 export 导出了 visibilityFilter 模块。

15.10 任务管理器应用测试

本节我们测试运行一下任务管理器应用，页面初始效果如图 15.3 所示。

图 15.3　任务管理器应用初始页面

然后，就可以在文本输入框中依次输入一组任务名称，通过"新增任务项"按钮添加进去，效果如图 15.4 和图 15.5 所示。

图 15.4　任务管理器应用"新增任务项"（一）

图 15.5 任务管理器应用"新增任务项"（二）

然后，我们可以通过点击具体的列表项（例如，Internet Explorer、Microsoft Edge 和 Safari），将不想占用内存的任务项禁用，被点击的任务列表项上会显示一条删除线，效果如图 15.6 所示。

图 15.6 任务管理器应用"禁用操作"（三）

下面，我们测试"全部""激活"和"禁用"这几个筛选项，效果如图 15.7、图 15.8 和图 15.9 所示。

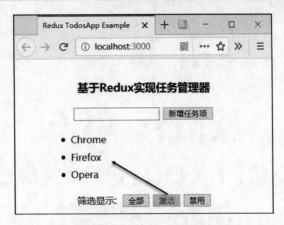

图 15.7 任务管理器应用"激活"筛选

图 15.8 任务管理器应用"禁用"筛选

图 15.9 任务管理器应用"全部"筛选

如图 15.7、图 15.8 和图 15.9 所示，通过"全部""激活"和"禁用"这几个筛选功能，可以选择出相对应的任务列表项。

第 16 章

实战5：基于 React+Router+Redux 的网站架构

本章介绍一个基于 React + Router + Redux 所实现的网站架构应用，其中 Router 就是指 React-Router 路由功能。

16.1 设计思想

本章介绍的这个网站架构应用非常简单，主要设计了一个"登录"模块、一个"主页"模块和一个"关于"模块。这个网站虽然简单，但使用到的几项关键技术都是本书的重点内容。比如：网站页面主要使用 React 框架技术，页面路由主要使用 React-Router 技术（模拟单页面效果），以及管理关键数据主要使用 Redux 技术。因此，本章所介绍的内容算是一个总结报告。

为了便于读者更好地理解这个示例，我们根据网站架构应用的主体架构绘制了一张逻辑图，如图 16.1 所示。

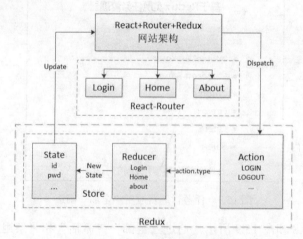

图 16.1　网站应用逻辑图

如图 16.1 所示,我们主要设计了"Login""Home"和"About"三个模块,其中"Login"是登录模块。

网站架构应用的代码结构如图 16.2 所示。

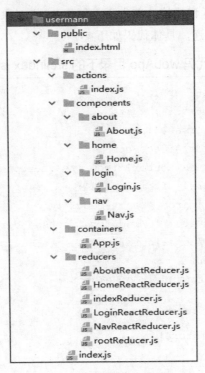

图 16.2 网站架构应用结构图

16.2 网站架构应用页面

首先,看一下网站架构应用的 HTML 主页,具体代码如下:

【代码 16-1】(详见源代码 webApp 目录下的 public\index.html 文件)

```
01  <!DOCTYPE html>
02  <html lang="en">
03    <head>
04      <meta charset="utf-8" />
05      <meta name="viewport" content="width=device-width, initial-scale=1" />
06      <title>React+Router+Redux webApp</title>
07    </head>
08    <body>
09      <div id="root"></div>
10    </body>
11  </html>
```

16.3 主入口模块

对于网站架构应用的主入口模块，首先还是创建 Store 对象，然后通过 Provider 容器组件方式引入 Store 对象并向下传递，具体代码如下：

【代码 16-2】（详见源代码 webApp 目录下的 src\index.js 文件）

```
01  import 'core-js/es6/map';
02  import 'core-js/es6/set';
03  import React from 'react';
04  import ReactDOM from 'react-dom';
05  import {HashRouter as Router} from 'react-router-dom';
06  import { createStore } from 'redux';
07  import { Provider } from 'react-redux';
08  import rootReducer from './reducers/rootReducer.js'
09  import App from './containers/App.js';
10  // TODO: createStore
11  const store = createStore(rootReducer);
12  // TODO: render
13  ReactDOM.render(
14    <Provider store={store}>
15      <Router>
16        <App />
17      </Router>
18    </Provider>
19    , document.getElementById('root')
20  );
```

关于【代码 16-2】的说明：

- 第 11 行代码通过 createStore() 创建了一个 store 对象。
- 第 14~18 行代码通过 <Provider> 定义了一个容器组件，同时添加了 store 属性。其中，比较特别的是第 15~17 行代码定义的 <Router> 路由模块，第 16 行代码在 <Router> 路由模块中包裹了一个 <App> 模块组件。

16.4 App 组件

在【代码 16-2】中，第 16 行代码引用的 <App> 模块组件是负责定义页面组件的，具体代码如下：

【代码 16-3】（详见源代码 webApp 目录下的 src\containers\App.js 文件）

```
01  import React, { Component } from 'react';
02  import {
03    Route,
04    Link,
05    Switch
06  } from 'react-router-dom';
07  import NavReactReducer from '../reducers/NavReactReducer.js';
08  import LoginReactReducer from '../reducers/LoginReactReducer.js';
09  import HomeReactReducer from '../reducers/HomeReactReducer.js';
10  import AboutReactReducer from '../reducers/AboutReactReducer.js';
11  // TODO: App Component
12  class App extends Component {
13    render() {
14      return (
15        <div>
16          <NavReactReducer /><br/><br/>
17          <Switch>
18            <Route exact path="/" component={LoginReactReducer}/>
19            <Route exact path="/Home" component={HomeReactReducer}/>
20            <Route path="/About" component={AboutReactReducer}/>
21          </Switch>
22        </div>
23      );
24    }
25  }
26  // TODO: export App
27  export default App;
```

关于【代码 16-3】的说明：

- 第 02～06 行代码引入了 React-Router 模块（Route、Link 和 Switch）。
- 第 07～10 行代码引入了一组自定义的 Reducer 模块。
- 第 12～25 行代码定义的 App 组件，包含了 1 个页面导航 Reducer 组件（NavReactReducer）以及一组路由组件，具体说明如下：
  - 第 16 行代码引入了一个<NavReactReducer>导航组件，用于页面导航的操作。
  - 第 18～20 行代码通过<Switch>和<Route>标签分别引入了<LoginReactReducer>子组件、<HomeReactReducer>子组件和<AboutReactReducer>子组件，用于页面展示操作。
- 第 27 行代码通过 export 导出 App 组件。

16.5 Reducer 设计

下面，我们看一下【代码 16-2】和【代码 16-3】中所引入的相关 Reducer 组件，具体代码如下：

【代码 16-4】（详见源代码 webApp 目录下的 src\reducers\rootReducer.js 文件）

```
01  import { combineReducers } from 'redux';
02  // 全局 reducer
03  import isLogin from './indexReducer.js'
04  // 合并 reducer
05  var rootReducer = combineReducers({
06    isLogin
07  });
08  // export rootReducer
09  export default rootReducer;
```

关于【代码 16-4】的说明：

- 第 01 行代码从 redux 引入了 combineReducers 方法。
- 第 02 行代码引入了一个子模块（isLogin），源文件名称为"indexReducer.js"。
- 第 05～09 行代码通过 combineReducers()方法包裹了上面引入的子模块（isLogin），然后通过 export 导出该 Reducer。

下面再看一下关于 isLogin 模块的代码：

【代码 16-5】（详见源代码 webApp 目录下的 src\reducers\indexReducer.js 文件）

```
01  import { UserMannType } from '../actions';
02  // TODO: reducer
03  var isLogin=false;
04  // TODO: indexReducer
05  function indexReducer(state = isLogin, action) {
06   switch (action.type) {
07      case UserMannType.LOG_IN:
08          // Login
09          return true;
10      case UserMannType.LOG_OUT:
11          // Logout
12          return false;
13      default:
14          return state;
15   }
16  }
```

```
17  // TODO: export indexReducer
18  export default indexReducer;
```

关于【代码 16-5】的说明：

- 第 03 行代码初始化了一个 State（isLogin=false），用于表示登录状态。
- 第 05～15 行代码定义了一个 Reducer 函数（indexReducer），通过判断 action 类型（LOG_IN 和 LOG_OUT）来执行页面操作。
- 第 18 行代码通过 export 导出了 indexReducer 模块。

下面再看一下关于 LoginReactReducer 模块的代码：

【代码 16-6】（详见源代码 webApp 目录下的 src\reducers\LoginReactReducer.js 文件）

```
01  import { connect } from 'react-redux';
02  import Login from '../components/login/Login.js';
03  import { UserMannType } from '../actions';
04  // TODO: mapStateToProps
05  function mapStateToProps(state) {
06    return {}
07  }
08  // TODO: mapDispatchToProps
09  function mapDispatchToProps(dispatch) {
10    return {
11      LOGIN: function(username, password, history) {
12        console.log("username: " + username);
13        console.log("password:" + password);
14        setTimeout(function() {
15          dispatch({type: UserMannType.LOG_IN});
16          history.push({pathname:'/Home'});
17        }, 1000);
18      }
19    };
20  }
21  // TODO: connect
22  var LoginReactReducer = connect(
23    mapStateToProps,
24    mapDispatchToProps
25  )(Login);
26  // TODO: export
27  export default LoginReactReducer;
```

关于【代码 16-6】的说明：

- 第 05～07 行代码定义的 mapStateToProps() 方法，负责将需要管理的 State（状态）映射到 Props 参数上。

- 第 09~20 行代码定义的 mapDispatchToProps()方法，负责将 dispatch()方法映射到 Props 参数上。其中，第 16 行代码通过调用 history 对象的 push()方法跳转到路径（'/Home'）。
- 第 22~27 行代码通过调用 connect()方法关联 Login 组件并导出该组件。

下面再看一下关于 HomeReactReducer 模块的代码：

【代码 16-7】（详见源代码 webApp 目录下的 src\reducers\HomeReactReducer.js 文件）

```
01  import { connect } from 'react-redux';
02  import Home from '../components/home/Home.js';
03  import { UserMannType } from '../actions';
04  // TODO: mapDispatchToProps
05  function mapStateToProps(state) {
06    return {
07        isLogin:state.isLogin
08    }
09  }
10  // TODO: mapDispatchToProps
11  function mapDispatchToProps(dispatch) {
12    return {
13        LOGOUT: function(history) {
14            dispatch({type: UserMannType.LOG_OUT});
15            history.push("/");
16        }
17    };
18  }
19  // TODO: connect
20  var HomeReactReducer = connect(
21    mapStateToProps,
22    mapDispatchToProps
23  )(Home);
24  // TODO: export
25  export default HomeReactReducer;
```

关于【代码 16-7】的说明：

- 第 05~08 行代码定义的 mapStateToProps()方法，负责将需要管理的 State（状态）映射到 Props 参数上。其中，第 07 行代码返回了用户登录状态（isLogin）。
- 第 11~18 行代码定义的 mapDispatchToProps()方法，负责将 dispatch()方法映射到 Props 参数上。其中，第 15 行代码通过调用 history 对象的 push()方法跳转到路径（'/'）。
- 第 20~25 行代码通过调用 connect()方法关联 Home 组件并导出该组件。

下面再看一下关于 NavReactReducer 模块的代码：

【代码16-8】（详见源代码 webApp 目录下的 src\reducers\NavReactReducer.js 文件）

```
01  import { connect } from 'react-redux';
02  import Nav from '../components/nav/Nav.js'
03  // TODO: mapStateToProps
04  function mapStateToProps(state) {
05    return {
06        isLogin:state.isLogin
07    }
08  }
09  // TODO: mapDispatchToProps
10  function mapDispatchToProps(dispatch) {
11    return {};
12  }
13  // TODO: connect
14  var NavReactReducer = connect(
15    mapStateToProps,
16    mapDispatchToProps
17  )(Nav);
18  // TODO: export
19  export default NavReactReducer;
```

关于【代码16-8】的说明：

- 第 04~08 行代码定义的 mapStateToProps()方法，负责将需要管理的 State（状态）映射到 Props 参数上。其中，第 06 行代码返回了用户登录状态（isLogin）。
- 第 10~12 行代码定义的 mapDispatchToProps()方法，负责将 dispatch()方法映射到 Props 参数上。
- 第 14~19 行代码通过调用 connect()方法关联 Nav 组件并导出该组件。

最后，还有一个比较简单的 AboutReactReducer 模块，读者可以自行查阅代码。

16.6 视图组件

在【代码16-6】中，第 25 行代码关联的<Login>组件负责用户登录操作，具体代码如下：

【代码16-9】（详见源代码 webApp 目录下的 src\components\login\Login.js 文件）

```
01  import React, { Component } from 'react';
02  // TODO: Login Component
03  class Login extends Component {
04    // TODO: render
05    render() {
```

```
06      return (
07        <div>
08          <h3>Login Page</h3>
09          <div>
10            Username:  <input type="text" ref="username" />
11          </div>
12          <div>
13            Password:  <input type="password" ref="password" />
14          </div>
15          <div>
16            <button onClick={this.userLogin.bind(this)}>Login</button>
17          </div>
18        </div>
19      );
20    }
21    // TODO: goLogin
22    userLogin() {
23      this.props.LOGIN(this.refs.username.value, this.refs.password.value, this.props.history);
24    }
25    // TODO: componentDidMount
26    componentDidMount() {
27      console.log("Login render complete.");
28    }
29  }
30  // TOOD: export Login
31  export default Login;
```

关于【代码 16-9】的说明：

- 第 03~29 行代码定义的 Login 组件，具体说明如下：
  - 第 07~18 行代码定义了一个用户登录功能（用户名输入框、密码输入框和提交按钮）。
  - 第 22~24 行代码定义了提交按钮的事件处理方法，通过 Props 参数调用"LOGIN"方法执行登录操作。
- 第 31 行代码通过 export 并导出了该组件。

在【代码 16-7】中，第 23 行代码关联的<Home>组件负责展示网站主页，具体代码如下：

【代码 16-10】（详见源代码 webApp 目录下的 src\components\home\Home.js 文件）

```
01  import React, { Component } from 'react';
02  import { Redirect } from 'react-router-dom';
03  // TODO: Home Component
04  class Home extends Component {
```

```
05        // TODO: render
06        render() {
07            if(this.props.isLogin==false){
08                return <Redirect to="/" />
09            }
10            return (
11                <div>
12                    <div>
13                        <button onClick={this.userLogout.bind(this)}>Logout</button>
14                    </div>
15                    <div>
16                        <h3>Home</h3>
17                        <p>Hi, this is home page.</p>
18                    </div>
19                </div>
20            );
21        }
22        // TODO: Logout
23        userLogout(){
24            this.props.LOGOUT(this.props.history);
25        }
26        // TODO: componentDidMount
27        componentDidMount() {
28            console.log("Home render complete.");
29        }
30    }
31    // TODO: export
32    export default Home;
```

关于【代码 16-10】的说明：

- 第 04～21 行代码定义的 Home 组件，具体说明如下：
  - 第 11～19 行代码定义了一个用户登出功能按钮和一些简单的主页内容。
  - 第 23～25 行代码定义了登出按钮的事件处理方法，通过 Props 参数调用 "LOGOUT" 方法执行登出操作。
- 第 32 行代码通过 export 并导出了该组件。

在【代码 16-8】中，第 17 行代码关联的<Nav>组件负责展示页面导航链接，具体代码如下：

【代码 16-11】（详见源代码 webApp 目录下的 src\components\nav\Nav.js 文件）

```
01    import React, { Component } from 'react';
02    import {
```

```
03    Route,
04    Link
05  } from 'react-router-dom';
06  // TODO: Nav Component
07  class Nav extends Component {
08    render() {
09      return (
10        <ul style={{display:this.props.isLogin?"block":"none"}}>
11          <li style={{display:this.props.isLogin?"none":"block"}}>
12              <Link to="/">Login</Link>
13          </li>
14          <li>
15              <Link to="/Home">Home</Link>
16          </li>
17          <li>
18              <Link to="/About">About Us</Link>
19          </li>
20        </ul>
21      );
22    }
23  }
24  // TODO: export Nav
25  export default Nav;
```

关于【代码 16-11】的说明：

- 第 07～23 行代码定义的 Nav 组件，具体说明如下：
  ➢ 第 10～20 行代码通过<ul><li>标签元素定义了一个导航链接，通过 Props 参数（isLogin）来判断用户登录状态，以选择导航链接的展示状态。
- 第 25 行代码通过 export 并导出了该组件。

最后，还有一个比较简单的 About 模块，读者可以自行查阅代码。

16.7 Action 定义

本节介绍网站架构应用的 Action 定义，具体代码如下：

【代码 16-12】（详见源代码 webApp 目录下的 src\actions\index.js 文件）

```
01  export const UserMannType = {
02    LOG_IN: 'LOG_IN',
03    LOG_OUT: 'LOG_OUT'
04  };
```

关于【代码 16-12】的说明：

- 第 02 行代码定义的 Action 类型（LOG_IN）表示用户登录操作。
- 第 03 行代码定义的 Action 类型（LOG_OUT）表示用户登出操作。

16.8 网站架构应用测试

本节测试运行网站架构应用，页面初始效果是一个登录界面，如图 16.3 所示。然后，可以模拟登录操作（未实现真正的登录验证功能），页面效果如图 16.4 和图 16.5 所示。

点击"About Us"链接测试 React Router 路由操作，效果如图 16.6 和图 16.7 所示。我们测试登出"Logout"功能，效果如图 16.8 所示。

图 16.3　网站架构应用初始登录界面

图 16.4　登录操作

图 16.5　网站主页

图 16.6　路由测试（一）

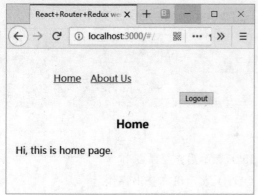

图 16.7 路由测试（二）

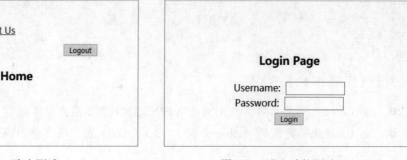

图 16.8 登出功能测试